GW01481784

Energy and Environment Research in China

More information about this series at http://www.springer.com/series/11888

Jiang Wu · Jianxing Ren
Weiguo Pan · Ping Lu · Yongfeng Qi

Photo-catalytic Control Technologies of Flue Gas Pollutants

Jiang Wu
School of Energy and Mechanical
  Engineering
Shanghai University of Electric Power
Shanghai, China

Jianxing Ren
Shanghai University of Electric Power
Shanghai, China

Weiguo Pan
Shanghai University of Electric Power
Shanghai, China

Ping Lu
School of Energy and Mechanical
  Engineering
Nanjing Normal University
Nanjing, China

Yongfeng Qi
School of Hydraulic, Energy and Power
  Engineering
Yangzhou University
Yangzhou, Jiangsu, China

ISSN 2197-0238　　　　　ISSN 2197-0246　(electronic)
Energy and Environment Research in China
ISBN 978-981-10-8748-6　　　ISBN 978-981-10-8750-9　(eBook)
https://doi.org/10.1007/978-981-10-8750-9

Jointly published with Shanghai Jiao Tong University Press, Shanghai, China

The print edition is not for sale in China Mainland. Customers from China Mainland please order the print book from: Shanghai Jiao Tong University Press.

Library of Congress Control Number: 2018954604

© Shanghai Jiao Tong University Press and Springer Nature Singapore Pte Ltd. 2019
This work is subject to copyright. All rights are reserved by the Publishers, whether the whole or part of the material is concerned, specifically the rights of translation, reprinting, reuse of illustrations, recitation, broadcasting, reproduction on microfilms or in any other physical way, and transmission or information storage and retrieval, electronic adaptation, computer software, or by similar or dissimilar methodology now known or hereafter developed.
The use of general descriptive names, registered names, trademarks, service marks, etc. in this publication does not imply, even in the absence of a specific statement, that such names are exempt from the relevant protective laws and regulations and therefore free for general use.
The publishers, the authors and the editors are safe to assume that the advice and information in this book are believed to be true and accurate at the date of publication. Neither the publishers nor the authors or the editors give a warranty, express or implied, with respect to the material contained herein or for any errors or omissions that may have been made. The publishers remains neutral with regard to jurisdictional claims in published maps and institutional affiliations.

This Springer imprint is published by the registered company Springer Nature Singapore Pte Ltd.
The registered company address is: 152 Beach Road, #21-01/04 Gateway East, Singapore 189721, Singapore

Preface

Since 1972, Fujishima and Honda found that titanium dioxide semiconductor electrode can be used to decompose water to hydrogen, and experts in chemistry, physics, chemical engineering and materials science have continued to study and understand the basic process of semiconductor photocatalytic materials and developed lots of new photocatalytic materials. In recent years, photocatalysis in environmental research and application has become one of the most active areas, and it has achieved encouraging results in practical application. But the application in the power plant emissions of pollutants is still in the blank.

The pollutants such as SO_2, NO_x and Hg emission from the coal-fired power station are harmful to environment and human health, so its emission controls have caused much concern in this world. There are various ways to reduce pollutants yielded in coal combustion power station, such as sorbent injection, catalytic oxidation and air pollution control devices. However, photocatalytic technology to reduce SO_2, NO_x and Hg from flue gas is a new technology. Compared with other methods, photocatalytic oxidization possesses higher oxidation ability and is no secondary pollution, therefore it is a promising technology to remove pollutants from the flue gas.

There are various kinds of factors affecting the mercury removal efficiency of the photocatalytic oxidation, such as the light sources, i.e., different wavelengths of the light sources, light intensity, reaction temperature, the characteristics of photocatalysts, and so on. We have designed different experiments to study the effects of these factors on photocatalytic performance.

We have studied titanium-based, zinc-based and bismuth-based photocatalysts in this book. The different morphologies and different doped items have the influence on the photocatalyst. In order to explore the influence of morphology, we prepared areatus spherical titanium dioxide photocatalyst. To explore the influence of metallic oxide, we prepared V_2O_5/titanium dioxide photocatalysts. To observe the co-influence of metallic oxide and nonmetal, we prepared carbon spheres supported CuO/titanium dioxide photocatalysts and carbon decorated In_2O_3/titanium dioxide photocatalysts. Moreover, zinc-based and bismuth-based photocatalysts were fabricated by different methods, such as hydrothermal method, coprecipitation method and calcination method.

The phase identification of as-prepared samples was characterized by X-ray diffraction (XRD), surface structure and morphology by scanning electron microscope (SEM), the shape and microstructure by transmission electron microscopy (TEM), the crystalline structure by HRTEM (high resolution TEM), the composition by the energy-dispersive x-ray spectroscopy (EDS), and optical absorption properties by the UV-vis diffuse reflectance spectra (UV-vis DRS). The specific surface area of the as-prepared photocatalysts was determined from the N_2 adsorption/desorption data, and photoluminescence (PL) spectra. Fourier transform infrared (FTIR) spectra and X-ray photoelectron spectroscopy (XPS) analysis were adopted to further identify the elemental composition.

A photocatalytic reactor was designed and adopted to evaluate the photocatalytic oxidation ability of different photocatalysts on elemental mercury in the simulated flue gas. After that, the effect of photocatalysts on the SO_2, NO_x, HCl, CO_2, H_2O, and Hg removal in the flue gas was studied. The relative mechanism on photoinduced electron-hole pair and its recombination, electron transfer and photocatalytic reaction have been studied. Photocatalytic clean technology can be applied not only in the gas pollutant cleaning at power plant but also in the purification of waste-water treatment.

This book presents nano-photocatalysis in two aspects of theory and application, including the concept of photocatalyst, photocatalytic reaction mechanism and kinetics, and photocatalytic reactor design. At the same time, the photocatalytic application in the control of flue gas pollutants was observed in detail. The readers may attain comprehensive understanding of possible mercury emission control methods and may put some technology into industrial application.

The presented work in this book is partly partially sponsored by National Natural Science Foundation of China (21237003, 50806041), and Natural Science Foundation of Shanghai (18ZR1416200).

Shanghai, China	Prof. Dr. Jiang Wu
Shanghai, China	Prof. Dr. Jianxing Ren
Shanghai, China	Prof. Dr. Weiguo Pan
Nanjing, China	Prof. Dr. Ping Lu
Yangzhou, China	Dr. Yongfeng Qi

Contents

1 **Foundations of Photocatalytic** 1
   1.1 History of Photocatalytic 1
   1.2 Basic Concepts of Photocatalytic 4
   1.3 Photochemical Reaction Principles 5
   1.4 Laws of Photochemistry 6
   1.5 Photocatalytic Reaction Theory of Semiconductor 8
   References 9

2 **Preparation and Characterization of Titanium-Based Photocatalysts** 13
   2.1 Preparation of Titanium-Based Photocatalysts 14
      2.1.1 Preparation of V_2O_5/Titanium Dioxide Photocatalysts 14
      2.1.2 Preparation of Carbon Spheres Supported CuO/Titanium Dioxide Photocatalysts 14
      2.1.3 Preparation of Carbon Decorated In_2O_3/Titanium Dioxide Photocatalysts 15
   2.2 Characterization of Titanium-Based Photocatalysts 15
      2.2.1 Characterization of V_2O_5/Titanium Dioxide Photocatalysts 15
      2.2.2 Characterization of Carbon Spheres Supported CuO/Titanium Dioxide Photocatalysts 26
      2.2.3 Characterization of Carbon Decorated In_2O_3/Titanium Dioxide Photocatalysts 34
   References 42

3 **Preparation and Characterization Other Photocatalysts** 45
   3.1 Zinc-Based Photocatalysts 45
   3.2 Bismuth-Based Photocatalysts 47

		3.2.1	Introduction of Bi-Based Photocatalysts	47

- 3.2.1 Introduction of Bi-Based Photocatalysts 47
- 3.2.2 Characterization of $BiOIO_3$ Nanosheets 49
- 3.2.3 Characterization of CSs-BiOI/$BiOIO_3$ Composites with Heterostructures 56
- References 63

4 Modified Photocatalysts 65
- 4.1 Morphology Controlled Photocatalyst Synthesis Methods 65
  - 4.1.1 Titanium Dioxide Hollow Microspheres Photocatalysts ... 65
  - 4.1.2 Anatase Titanium Dioxide with Co-exposed (001) and (101) Facets 66
- 4.2 Metal or Nonmetal Modified Zinc Base Photocatalysts 66
  - 4.2.1 Doping Metals 66
  - 4.2.2 Doping Nonmetals 69
- 4.3 Metal or Nonmetal Modified Titanium Dioxide Photocatalysts ... 70
  - 4.3.1 CuO/Titanium Dioxide Photocatalysts 70
  - 4.3.2 V_2O_5/Titanium Dioxide Photocatalysts 72
  - 4.3.3 Carbon Spheres Supported CuO/Titanium Dioxide Photocatalysts 73
  - 4.3.4 Carbon Decorated In_2O_3/Titanium Dioxide Photocatalysts 74
- 4.4 Metal or Nonmetal Modified $BiVO_4$ Photocatalysts 75
- 4.5 Graphene Supported Titanium Dioxide Photocatalysts 75
- References 76

5 Photocatalytic Denitrification in Flue Gas 83
- 5.1 Denitrification in the Flue Gas 83
  - 5.1.1 The Importance of Denitration 83
  - 5.1.2 Photocatalytic Technology 85
- 5.2 Photocatalytic Denitrification in Flue Gas 88
  - 5.2.1 Experimental 89
  - 5.2.2 Results and Discussion 91
  - 5.2.3 Mechanism of SO_2 Removal by TiO_2 Photocatalysis 95
  - 5.2.4 Mechanism of NO Removal by Titanium Dioxide Photocatalysis 96
- 5.3 Denitrification in Power Plant Flue Gas 96
  - 5.3.1 The Principle of SCR to Remove NO_x 96
  - 5.3.2 The Physical Shape Classification and Characteristics of SCR Catalyst 97
  - 5.3.3 The Chemical Material Classification of SCR Catalyst ... 98
  - 5.3.4 High-Temperature and Low-Temperature Catalysts 98
- References 100

6 The Photocatalytic Removal of Mercury from Coal-Fired Flue Gas ... 103
- 6.1 Measurement of Photoactivity ... 104
- 6.2 The Photocatalytic Removal of Mercury by Metal or Nonmetal Modified Titanium Dioxide Photocatalysts ... 105
  - 6.2.1 The Photocatalytic Removal of Mercury by V_2O_5/Titanium Dioxide Photocatalysts ... 105
  - 6.2.2 The Photocatalytic Removal of Mercury by Carbon Spheres Supported CuO/Titanium Dioxide Photocatalysts ... 114
  - 6.2.3 The Photocatalytic Removal of Mercury by Carbon Decorated In_2O_3/Titanium Dioxide Photocatalysts ... 116
- 6.3 The Photocatalytic Removal of Mercury by Other Photocatalysts ... 125
  - 6.3.1 The Photocatalytic Removal of Mercury by $BiOIO_3$ ... 125
  - 6.3.2 The Photocatalytic Removal of Mercury by CSs-BiOI/$BiOIO_3$ ... 129
  - 6.3.3 The Photocatalytic Removal of Mercury by ZnO ... 134
- References ... 138

7 The Photocatalytic Technology for Wastewater Treatment ... 141
- 7.1 The Principle of Photocatalytic Wastewater Treatment ... 141
- 7.2 Titanium-Based Photocatalysts ... 142
  - 7.2.1 Pure Titanium Dioxide ... 142
  - 7.2.2 Tungsten-Doped Titanium Dioxide ... 143
  - 7.2.3 Ag^+-Doped Titanium Dioxide ... 145
- 7.3 Other Photocatalysts Used in Wastewater Treatment ... 147
  - 7.3.1 Zinc-Based Photocatalysts ... 147
  - 7.3.2 Bismuth-Based Photocatalysts ... 148
- References ... 148

Index ... 151

Chapter 1
Foundations of Photocatalytic

Abstract Semiconductor photocatalysis technology is a green technology being developed rapidly in recent years by using solar energy for energy conversion and environmental purification, which has important application in both energy and environment. Photocatalysis has gradually been considered as a promising way to solve energy and environmental problems since Fujishima achieved hydrogen production from cracking water using titanium dioxide (titanium dioxide) in 1972. This field is becoming a hot topic and has attracted a lot of scientists' attention and research. Semiconductor catalyst is driven by light to convert light energy into electrical or chemical energy without additional pollution. Furthermore, it is a technology to produce environmental friendly clean energy only with sunlight and can solve the ever increasing environmental pollution problems. In recent years, the depletion of fossil fuel and requirement of clean environment push the development of renewable energy including solar energy, which provides a good platform for the photocatalytic technology to accomplish the dream of blue sky and white cloud and sustainable developing. This chapter focuses on the development of photocatalytic technology for decades and expounds its development from the advent to the prosperous.

Keywords Photocatalytic · Titanium dioxide · Nanoparticles · Principles Photochemistry

1.1 History of Photocatalytic

The photocatalyst was found in 1967 by Professor Fujishima Akira, who was a graduate student at the time in the University of Tokyo. The single crystal of titanium oxide in water was irradiated with light in one experiment, and it came to the end that water was decomposed into oxygen and hydrogen. This is called 'Honda-Fujishima Effect' [1]. It combines the name of Professor Fujishima Akira and his advisor, Honda Kenichi, President of Tokyo Technology University. In a popular sense, catalyst is the meaning of catalyst as its name implies. The catalyst is a chemical substance that accelerates chemical reactions without itself undergoing any change. Photocatalyst is a general term for the chemical substances that can act as a catalyst by the excitation of photons [2]. Photocatalytic technology is the basic nanotechnology that was founded in 1970s. In the mainland of China, we generally use the popular term as a photocatalyst to call the photocatalyst. The typical natural photocatalyst is our common chlorophyll, which promotes carbon dioxide and hydration in air to become oxygen and carbohydrates in plant photosynthesis. It can decompose all organic and inorganic substances that are harmful to the human body and the environment. Not only can it speed up reactions, but it can also use natural laws without wasting resources and creating additional pollution [3]. The most representative example is the 'Photosynthesis' of plants: they absorb carbon dioxide, which is harmful to animals, and convert to oxygen and water by using light energy.

In 1976, John H Carey studied the photocatalytic oxidation of polychlorinated biphenyls, which is considered as the pioneering research work of photocatalytic technology in eliminating environmental pollutants. In 1977, Yokota T found that titanium dioxide had photocatalytic activity for propylene epoxidation under light conditions, which widened the application scope of photocatalysis and provided a new way for organic oxidation. Since 1983, A.L. Pruden and D. Follio have continuously studied the photocatalytic oxidation of a series of pollutants such as alkanes, olefins and aromatic hydrocarbons. It is found that all reactants can rapidly degrade. In 1989, Tanaka K found that the organic semiconductor photocatalytic process was induced by hydroxyl radical (OH). Adding H_2O_2 to the system could increase the concentration of OH^- [4]. In the 1990s, with the rapid development of nanotechnology and photocatalysis technology in environmental protection, health care, organic synthesis and other applications, the research of nanometer scale photocatalyst has been proved to be one of the most animate fields of investigation around the world.

In 1992, the first International Symposium on titanium dioxide photocatalyst was held in Canada. Japan's research institutes published many new ideas about photocatalyst, and put forward the research achievements applied to nitrogen oxides purification. Therefore, the number of patents related to titanium dioxide is also the largest. Other catalyst-related technologies include catalyst preparation, catalyst structure, catalyst support, catalyst fixation, catalyst performance testing and so on. Taking this opportunity, the research on photocatalyst in the field of antibacterial, antifouling and air purification has increased dramatically. From 1971 to June 2000, there were 10,717 patents related to photocatalyst. The wide application of titanium dioxide titanium dioxide photocatalyst will bring a clean environment and healthy body for people. During the 1975–1985, artificial photosynthesis and solar energy conversion have been regarded as one of the 14th largest projects in China in the National Compendium on Basic Scientific Development. Photocatalytic decomposition of water to produce hydrogen is an effective way to utilize solar energy, which has developed rapidly in China. Because of its simplicity, pollution-free and easy operation, it has attracted extensive attention from some universities and research institutions, such as Dalian Institute of Chemical Physics of Chinese Academy of Sciences, Lanzhou Institute of Chemical Physics of Chinese Academy of Sciences, Jilin University, Shandong University, Wuhan University and so on. Generally, the water systems of photocatalytic decomposition were as below: those which use metal complexes as photosensitizers, and those which utilize semiconductor nanoparticles to constitute heterogeneous systems.

The main system of photocatalytic decomposition of water includes photosensitizer (Ru or Rh complex) with metal complexes, electron mediator, electron donor (TEOA or EDTA), surfactant, and other additives [5, 6]. When using metalloporphyrin and active group as photosensitizer, it can not only increase the efficiency of water splitting, but also improve stability and repeatability by preventing reverse reaction.

In Switzerland, the same phenomenon occurred in research system of developing $Ru(bpy)_3^{2+}$ by Grätzel and co-researchers. Moreover, semiconductor nanoparticles, easily amplified by the thin film design of photocatalytic layer, were also used for photocatalytic water decomposition and organic synthesis. Hydrogen and ethylene glycol produced by photoinduced dehydrogenation C-C coupling reaction were also reported with powder photocatalytic systems including ZnS, methanol, and water [7]. The high rate of evolution of hydrogen and the high selectivity of ethylene glycol (95%) approaching the advanced international standards were very important [7]. In the middle of 1980s, the photocatalytic decomposition of water reached the lowest point, and the photocatalytic degradation of pollutants and the photocatalytic reduction of heavy metals reached the peak. With the increasing environmental problems caused by economic growth, the photocatalysis of the environment is more popular than the photocatalysis of energy. The cleavage of sulfide was achieved into a semiconductor nanoparticles suspension. CdS was widely applied to be a representative photocatalyst for the organic pollution degradation. On the surface of CdS, photo deposition of RuO_2 or Rh_2O_3 was an effective method to avoid the slight corrosion behavior of CdS by rapid separation and transfer of electrons and holes induced by light. Thus, it prevented the charge on the surface area of the catalyst. Under visible light irradiation, the photocatalytic system consisting of Rh_2O_3/CdS exhibited activity in H_2S or sulfide cracking into hydrogen and sulfur. The reflectance spectra of CdS and Rh_2O_3/CdS confirmed that the Rh_2O_3 deposited on CdS particles promoted the absorption of light in the visible region. The photocatalytic technology had been explored to solve the problem of hydrogen sulfide pollution produced by tanneries, paper mills, oil refineries, chemical plants, and coal plants [8–10].

In 1995–2012 years, titanium dioxide based semiconductor is the most important catalyst in the first generation of photocatalyst. Its synthesis methods include sol-gel method, micelle method and reverse micelle method, hydrothermal method, solvothermal method, direct oxidation method etc. [11, 12]. With the research on the preparation and application of titanium dioxide materials, many related comments had emerged [13]. Because of its many advantages, titanium dioxide photocatalyst has become one of the useful photocatalyst materials. Its advantages as a photocatalyst include water insolubility, hydrophilicity, non-toxicity, cheapness and easy availability, biological compatibility, photo stability and having the proper band potentials and so on. In particular, the oxidative decomposition of organic pollutants into CO_2 and water. As the most famous photocatalyst, titanium dioxide plays an important role in solving the increasing environmental challenges. It provides great hope to overcome the energy crisis through the effective use of solar photovoltaic (such as dye sensitized solar cells). Titanium dioxide can be made into an intelligent surface with antibacterial, antifogging and self-cleaning properties by the construction of photocatalytic functional coatings and films on other materials. Currently, more and more photocatalytic functional coatings are being commercialized.

1.2 Basic Concepts of Photocatalytic

Most of the semiconductor photocatalyst is n type semiconductor materials (currently the most widely used are thought titanium dioxide), which is different from the metal or insulating material, especially the band structure, namely in the valence band (VB) and conduction band (CB) [14], and there is a gap (Forbidden Band, Band Gap) between them. Due to the relationship between the optical absorption threshold and the band gap of the semiconductor with the $k = 1240/E.g.$ (eV) relationship, the absorption wavelength threshold of the commonly used broadband gap semiconductor is mostly in the ultraviolet region. When the photon energy is higher than that of the semiconductor absorption threshold, the valence band electrons take place inter band transitions, that is, the transition from the valence band to the conduction band, resulting in the production of photo generated electrons (e^-) and holes (h^+) [15]. The adsorption of dissolved oxygen in the surface of nanoparticle trapping electrons to form superoxide anion, and the hole will be adsorbed on the catalyst surface hydroxyl ions and water oxidation of hydroxyl radicals. The superoxide anion and hydroxyl radical are highly oxidizing, which can oxidize most of the organic compounds to the final products CO_2 and H_2O, and even decompose some of the inorganic matters. Under the light excitation, the electrons move from the valence band to the guide band, thus forming the photo generated electrons in the conduction band and forming a light hole on the valence band [16]. By using the reduction and oxidation properties of the photo generated electron-hole pairs, the organic pollutants in the surrounding environment can be purified and the photolysis water can be adopted for the preparation of H_2 and O_2. High efficiency catalyst must meet the following conditions:

1. Semiconductor proper conduction band and valence band location.
   In the application of purifying pollutants, the valence band potential must have enough oxidation performance. In the application of photolysis water, the potential must meet the requirements of H_2 and products.
2. High efficiency electron-hole separation ability and reduce their compound probability.
   The photocatalyst is a solid semiconductor, and the reaction medium is usually gas or liquid, so often the photocatalytic process is called heterogeneous photocatalysis or semiconductor photocatalysis. More precisely, in accordance with the photocatalyst and photocatalytic reactions of different phases [17], the photocatalysis may be divided into solid-gas and liquid-solid photocatalysis and photocatalytic, and the photocatalysis occurring in the gas-solid surface is called

surface photocatalysis, and the other one is called interface photocatalysis due to the occurrence of the solid-liquid interface and interface photocatalytic reaction mechanism. These two kinds of situations are different: in the solid-gas photocatalysis process, oxygen is the oxidant precursor; while in the liquid-solid photocatalysis process, water molecular is the main oxidant precursor. According to the difference of primary process of photocatalytic reaction, the photocatalytic reaction of Linsebigler et al. are divided into two categories [18] said a sensitized photochemical reaction (sensitized catalyst photoreaction, excitation light generated electrons and holes, the electron and hole transfer to the ground and lead to the decomposition of adsorbed molecules; the A class called catalytic photochemical (reaction catalyzed photoreaction) optical excitation of adsorbed molecules, excited state molecules and ground catalyst interactions, the catalyst for the adsorbed molecules provides is advantageous to the reaction surface structure. By sound that [19, 20] recently, in some systems A and B two kinds of photocatalytic process may occur at the same time, there is a kind of C sensitized photocatalytic reaction.

1.3 Photochemical Reaction Principles

At present, photochemical reactions have attracted wide attention due to their excellent properties and wide application [21–24]. Photochemical reactions have special significance for thermal reactions, which are based on two major features of photochemical reactions. The reaction occurs in the excited state of a molecule with amount of energy compared with the ground state, and it is usually possible to achieve a thermodynamic disadvantageous reaction due to its ground state reactants. Second, reactions in general occur at low temperatures, so that products can be generated in cold state. Therefore, it is usually possible to overcome the high strain ring system by extracting excess energy as light to surmount the activation energy barrier in formation. The photochemical reactions are ordinarily carried out by irradiation with ultra-violet ($\lambda = 200$–350 nm) and are not often used in visible light. Consequently, these reactions usually require at least one kind of unsaturated or aromatic. Since common borosilicate Pyrex glass or sodium silicate glass vessels can only emit radiation greater than 300 nm and absorb less than this wavelength, these reactions need to be performed in a pure fused silica vessel with the aim of emitting UV radiation to meet the reaction criteria. The photochemical reaction of compounds is a chemical reaction caused by the interaction of molecules with ultraviolet or visible light. In photochemical reactions, the use of specific wavelengths of light can stimulate molecules. According to quantum theory, matter and light are quantized. Only a specific light energy can be stimulated by the absorption of specific organic molecules. The absorption or emission of light occurs through the transmission of energy by photons. These photons have both wave and particle properties at the same time. Photon energy E is given by Planck's Law (Eq. 1.1) [25]:

$$E = h\nu \tag{1.1}$$

where h is Planck constant and is equal to 6.63×10^{-34} Js and ν is the frequency of oscillation of the photon in units of s^{-1} or Hertz (Hz).

$$\nu = c/\lambda \tag{1.2}$$

where c is the velocity of light and λ is the wavelength of oscillation of photon. Thus,

$$E = h\nu = hc/\lambda \tag{1.3}$$

Therefore, the energy of a photon is directly proportional to its frequency and inversely proportional to its wave length. The energy of one mole of photons (6.02×10^{23} photons) is called an Einstein and is measured in units in kJ/mol. It is equal to Nhc/λ, in which N is the number of photons.

1.4 Laws of Photochemistry

The two basic principles related to the absorption of light by organic molecules are the basis for understanding its photochemical transformation.

1. The Grotthuss-Draper law:
   The law states that only the fraction of light which is absorbed by a chemical entity can bring about the photochemical change [26].
2. The Stark-Einstein law:
   The law states that each molecule or atom absorbs one photon or one quantum of light for its excitation or activation [27], i.e., for a molecule, AB,

$$AB(\text{ground state}) + h\nu \rightarrow AB^*(\text{excited state}) \tag{1.4}$$

   This law is consistent with most cases, but there would be exceptions at the time of irradiating samples with laser and other strong light sources.
3. The Beer-Lambert's Law of Light Absorption
   The extent of light absorbed by a substance depends on its molar absorption coefficient (ε). The fraction of light absorbed (I/I_0) by a substance is given by the Beer-Lambert law [28]. The law states that the ratio of the intensity of the emergent light (I) to incident light (I_0) has an exponential relationship with the concentration (c) and path length (l) of the absorbing substance, i.e.,

$$\frac{I}{I_0} = 10^{-\varepsilon c l} \tag{1.5}$$

   Taking logarithm to the base 10 gives

1.4 Laws of Photochemistry

$$\log\left(\frac{I}{I_0}\right) = -\varepsilon c l \tag{1.6}$$

The left-hand quantity is the absorbance, A, and hence

$$A = \varepsilon c l \tag{1.7}$$

where c is the concentration of the substance in moles per liter, mol L^{-1}, and l is the path length in cm.

The higher the ε value, the higher the absorption strength. Generally speaking, the optical absorption intensity of the optical transition process is high, and the light absorption intensity of the transition process is not allowed to be low.

4. Physical Basis of Light Absorption by Molecules:
The Franck-Condon Principle Chromophores or chromophoric groups present in the molecules are responsible for the absorption of light. The absorption of UV or visible light results in the promotion of electrons from their ground state orbit to highly excited states. Normally, the energy required for such a transformation depends mainly on the nature of the two tracks involved, while the rest of the molecule is less. When this electron transition occurs, the absorbing chromophores undergo electron dipole transitions. The transition dipole moment only lasts for a long time, and the absorption intensity of light is directly proportional to the square of the transition dipole moment. The total energy of a molecule consists of its electron energy and the energy of its core (vibration and rotation). It is expressed as

$$E_t = E_e + E_v + E_r \tag{1.8}$$

where the subscripts refer to the total energy, electronic energy, vibrational energy, and rotational energy respectively. The energy gap between the electronic states is much larger than that between the vibrational states, and the gap between the vibrational states is larger than that between the rotational states. The absorption of light by molecules causes the transition of electrons from one electronic state to another, which is faster than that of the nucleus, because the mass is heavier than the mass of the electrons. The electron transition takes place much faster than the nucleus of the vibrational molecule, so it can be considered that the nucleus is fixed during the electron transition cycle. Its name is the Franck-Condon principle [29]. It points out that the absorption of light by molecules will lead to electron transitions in stationary nuclear framework of the molecule.

Therefore, the electron transitions of absorbing photons are usually referred to as vertical transitions or Franck-Condon transitions. The electronic transitions in molecules by absorption of light lead to changes in the vibrational state of electrons. Thus, such electronic transition is called the vibrational transition of electrons. Figure 1.1 shows the potential energy curve known as Morse potential

Fig. 1.1 Schematic diagram of the electronic ground state and the first excited electronic state of a diatomic molecule

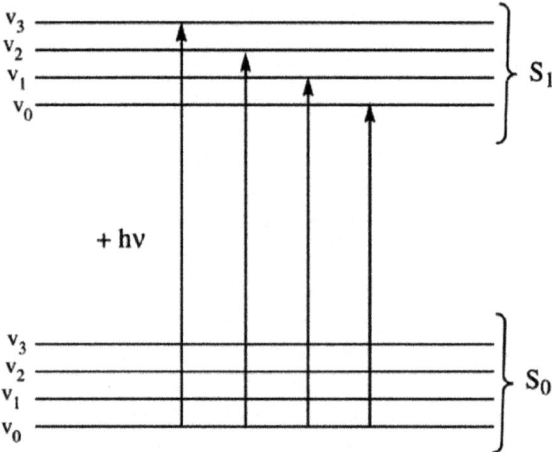

energy curve of a diatomic molecule for its electronic ground state (S_0) and the first excited electronic state (S_1).

1.5 Photocatalytic Reaction Theory of Semiconductor

Singlet and three excited states can undergo photochemical reactions. The lifetime of the singlet excited state is very short, and the lifetime of the three states is relatively long. Therefore, most photochemical reactions can be enhanced by the three excited states [30]. Excited molecules perform single molecular or bimolecular reactions in a single step (coordination process) or in two or multi-step processes involving one or more intermediates. Most photochemical reactions broke through photolysis and turned into free radicals coupling, isomerization, dimerization, hydrogen abstraction, elimination and rearrangements [31]. Absorption of a photon by an organic molecule, R, leads to the formation of an electronically excited state, R^* of the molecule.

$$R + h\nu \rightarrow R^* \quad (1.9)$$

The excited state R^* may react in any one of the two ways: in a concerted process (i.e., in a single step) gives the product P:

$$R^* \rightarrow P \quad (1.10)$$

These concerted processes include a series of pericyclic reactions from S1 (π, π^*) via cyclic transition states, where σ or π bonds are cleaved and formed simultaneously.

In two or multistep process, one or more intermediates I are formed:

$$R^* \rightarrow I \rightarrow P \tag{1.11}$$

or

$$R^* \rightarrow I_1 \rightarrow I_2 \rightarrow I_3, \text{etc} \rightarrow P \tag{1.12}$$

These include free radical reactions involving diradical intermediate or intermediates from either S1 (π, π^*; n, π^*) or T1 ($\pi \rightarrow \pi^*$, n, π^*). The most common photochemical reactions are reactions of carbonyl compounds, olefins, and aromatic compounds as well as chain reactions of hydrocarbons.

Semiconductor photocatalysis is a potential technology to support our future development [32]. As shown in Fig. 1.1, photocatalytic processes generally consist of three main steps: the first step is the optical absorption. A photocatalyst absorbs UV and/or visible (Vis) light irradiation from sunlight or an illuminant light source. The electrons in the valence band of the photocatalyst are excited to the conduction band, while the holes remain in the valence band. So it creates the negative-electron (e^-) and positive-hole (h^+) pairs [33]. The energy difference between valence band and conduction band is called "band gap". Then these light-induced electrons (or holes) need to first resist the compound induced by hole capture (or electron capture), and then migrate through the body and surface of the photocatalyst to reach the catalytic reaction site, in which the retained electrons and holes catalyze the decomposition of water. It can be used to reduce carbon dioxide and pollutant degradation [34]. In these photocatalytic processes, e^-/h^+ pairs form within several femtoseconds, their journeys from bulk to reactive sites need hundreds of picoseconds, and catalytic reactions between e^-/h^+ and adsorbed reactants occur within a few nanoseconds to a few microseconds [35]. On the contrary, electrons and holes recombine on a larger time scale that covers from several picoseconds to dozens of nanoseconds [36]. In particular, recombination in the bunk takes only a few picoseconds, much faster than charge transportation and charge-consuming catalysis [37]. This means that most photoelectrons and holes firstly recombine with each other in the photocatalyst bulk before reaching the surface of the photocatalyst. Consequently, subsequent surface migration and final surface catalysis only have less electrons and holes to utilize. Since this low volume charge separation efficiency leads to the poor performance of the photocatalyst in many studies, it is urgent to implement the large capacity of charge separation strategy for the development of semiconductor photocatalysis in the future.

References

1. T. Inoue et al., Photoelectrocatalytic reduction of carbon dioxide in aqueous suspensions of semiconductor powders. Nature **277**(5698), 637–638 (1979)
2. P. Wu, R. Xie, K. Imlay et al., Visible-light-induced bactericidal activity of titanium dioxide co-doped with nitrogen and silver. Environ. Sci. Technol. **44**(18), 6992–6997 (2010)

3. K.R. Reddy, M. Hassan, V.G. Gomes, Hybrid nanostructures based on titanium dioxide for enhanced photocatalysis. Appl. Catal. A Gen. **489**, 1–16 (2015)
4. D.O. Scanlon, C.W. Dunnill, J. Buckeridge et al., Band alignment of rutile and anatase TiO_2. Nat. Mater. **12**(9), 798–801 (2013)
5. G.X. Jin, W.Z. Hu, A mix model for calculating the wind speed frequency distribution. Acta Energiae Solaris Sinica, (1994)
6. N. Zhang et al., Synthesis of M@titanium dioxide (M = Au, Pd, Pt) core-shell nanocomposites with tunable photoreactivity. J. Phys. Chem. C **115**(18), 9136–9145 (2011)
7. Y. Lin, R. Lin, F. Yin et al., Photo electrochemical studies of H_2, evolution in aqueous methanol solution photocatalysed by Q-ZnS particles. J. Photochem. Photobiol. A Chem. **125**(1–3), 135–138 (1999)
8. Y. Li, F. Gao, W. Wei et al., Pore size of macroporous polystyrene microspheres affects lipase immobilization. J. Mol. Catal. B Enzym. **66**(1–2), 182–189 (2010)
9. Z.H. Wang, Q.X. Zhuang, Photocatalytic reduction of pollutant Hg (II) on doped WO_3, dispersion. J. Photochem. Photobiol. A Chem. **75**(2), 105–111 (1993)
10. D. Chen, A.K. Ray, Removal of toxic metal ions from wastewater by semiconductor photocatalysis. Chem. Eng. Sci. **56**(4), 1561–1570 (2001)
11. X. Zou, Y. Chen, X. Zhu et al., Preparation of the Keggin type chromium substituted phosphotungstates/titanium dioxide nano film and its visible photocatalytic performance. Appl. Chem. **33**(3), 320–329, in Chinese (2016)
12. Y.U. Quanwei, M. Zhao, Z. Liu et al., Catalytic decomposition of ozone in ground air by manganese-based monolith catalysts. Chin. J. Catal. **30**(1), 1–3 (2009)
13. L. Jing, Y. Qu, B. Wang et al., Review of photoluminescence performance of nano-sized semiconductor materials and its relationships with photocatalytic activity. Sol. Energy Mater. Sol. Cells **90**(12), 1773–1787 (2006)
14. X.S. Zhao, G.Q. Lu, G.J. Millar, Encapsulation of transition metal species into zeolites and molecular sieves as redox catalysts: part I-preparation and characterisation of nanosized titanium dioxide, CdO and ZnO semiconductor particles anchored in NaY zeolite. J. Porous Mater. **3**(1), 61–66 (1996)
15. H. Huang, X. Han, X. Li, S. Wang, P. K. Chu, Y. Zhang, Fabrication of multiple heterojunctions with tunable visible-light-active photocatalytic reactivity in BiOBr-BiOI full-range composites based on microstructure modulation and band structures. ACS Appl. Mater. Interfaces **7**(1), 482–492 (2015)
16. L. Xie, L. Ping, Z. Zheng et al., Morphology engineering of V_2O_5/titanium dioxide nanocomposites with enhanced visible light-driven photofunctions for arsenic removal. Appl. Catal. B **184**, 347–354 (2016)
17. D. Jiang, T. Wang, Q. Xua, D. Li, S. Meng, M. Chen, Perovskite oxide ultrathin nanosheets/g-C3N4 2D-2D heterojunction photocatalysts with significantly enhanced photocatalytic activity towards the photodegradation of tetracycline. Appl. Catal. B Environ. **201**, 617–628 (2017)
18. C. He, B. Shen, J. Chen et al., Adsorption and oxidation of elemental mercury over Ce-MnOx/Ti-PILCs.[J]. Environ. Sci. Technol. **48**(14), 7891–7898 (2014)
19. H. Huang, X. Li, J. Wang, F. Dong, P.K. Chu, T. Zhang, Y. Zhang, Anionic group self-doping as a promising strategy: band-gap engineering and multi-functional applications of high-performance C-doped $Bi_2O_2CO_3$. ACS Catal. **5**, 4094–4103 (2015)
20. G. Xiong, R. Shao, T.C. Droubay, A.G. Joly, K.M. Beck, S.A. Chambers, W.P. Hess, Photoemission electron microscopy of titanium dioxide anatase films embedded with rutile nanocrystals. Adv. Funct. Mater. **17**, 2133–2138 (2007)
21. V. Ramamurthy, K. Venkatesan, Photochemical reactions of organic crystals. Chem. Rev. **87**(2), 433–481 (1987)
22. F.H. Quina et al., Photochemical reactions in organized monolayer assemblies. Z. Phys. Chem. **101**(1–6), 151–162 (1976)
23. N. Hoffmann, ChemInform abstract: photochemical reactions as key steps in organic synthesis. Cheminform **39**(25), 1052–1103 (2008)

References

24. G. Sprintschnik, H.W. Sprintschnik, P.P. Kirsch, D.G. Whitten, Cheminform abstract: photochemical reactions in organized monolayer assemblies. 6. preparation and d photochemical reactivity of surfactant ruthenium(ii) complexes in monolayer assemblies and at water-solid interfaces. Am. Chem. Soc. **99**(15), 4947–4954 (1997)
25. S.B. Giddings, Hawking radiation, the Stefan-Boltzmann law, and unitarization. Phys. Lett. B **754**, 39–42 (2015)
26. J.W. Draper, *The Twelfth Night of Shakespeare's Audience [M]* (Octagon Books, 1975)
27. F.H. Thaheld, Can the Stark-Einstein law resolve the measurement problem from an animate perspective? Bio Systems **135**, 50 (2015)
28. T.C.E. Marcus et al., Alternative wavelength for linearity preservation of Beer-Lambert law in ozone concentration measurement. Microw. Opt. Technol. Lett. **57**(4), 1013–1016 (2015)
29. C. Noda, R.N. Zare, Relation between classical and quantum formulations of the Franck-Condon principle: the generalized r-centroid approximation. J. Mol. Spectrosc. **95**(2), 254–270 (1982)
30. W. Albert Noyes, Jr., G.S. Hammond, J.N. Pitts, Jr., *Properties and Reactions of Organic Molecules in Their Triplet States* (Wiley, 2007), pp. 21–156
31. M. Mansour, Photolysis of aromatic compounds in water in the presence of hydrogen peroxide. Bull. Environ. Contam. Toxicol. **34**(1), 89–95 (1985)
32. Y. Ma et al., Titanium dioxide-based nanomaterials for photocatalytic fue generations. Chem. Rev. **114**(19), 9987–10043 (2014)
33. S. Bai et al., Steering charge kinetics in photocatalysis: intersection of materials syntheses characterization techniques and theoretical simulations. Chem. Soc. Rev. **44**(10), 2893–2939 (2015)
34. K.R. Gopidas, P.V. Kamat, Photoinduced charge transfer processes in ultrasmall semiconductor clusters. Photophysical properties of CdS clusters in Nafion membrane. J. Chem. Sci. **105**(6), 505–512 (1993)
35. J.Z. Zhang, Interfacial charge carrier dynamics of colloidal semiconductor nanoparticles. J. Phys. Chem. B **104**(31), 7239–7253 (2000)
36. A. Kubacka, M. Fernández-García, G. Colón, Advanced nanoarchitectures for solar photocatalytic applications. Chem. Rev. **112**(3), 1555–1614 (2012)
37. J.F. Montoya et al., Comprehensive kinetic and mechanistic analysis of titanium dioxide photocatalytic reactions according to the direct-indirect model: (II) experimental validation. J. Phys. Chem. C **118**(26), 14276–14290 (2014)

Chapter 2
Preparation and Characterization of Titanium-Based Photocatalysts

Abstract We prepared a series of mole percentage of modified titanium dioxide materials. The crystalline phases of as-prepared nanocomposites were characterized by using the X-ray diffraction (XRD) of Cu Kα radiation (BRUKER D8 ADVANCE Diffractometer, Germany), in which the scanning angle range begins from 10° to 90° (the scanning rate is $2°\text{min}^{-1}$). In order to further identify the element composition, the AIK alpha X ray (HM = 1486.6 eV) radiation operated under 250 W (PHI5300, USA) was used to carry out the X ray photoelectron spectroscopy (XPS) analysis. In order to observe the surface structure and morphology, scanning electron microscope (SEM, Phillips XL-30 FEG/NEW) was used. Transmission electron microscopy (TEM, Philips Model CM200) confirmed the shape and microstructure. The crystal structures of the prepared samples were characterized by high resolution TEM (HRTEM). Energy dispersive X ray spectroscopy (EDS) was used to confirm the composition of the samples. According to the Brunauer-Emmett-Teller (BET) method (Micromeritics ASAP 2020), the error line of the specific surface area of the sample prepared from the N_2 adsorption/desorption data is 1% of the measurement result. Fourier transform infrared (FTIR) spectra were obtained by using Nicolet Nexus spectrometer in the range of 4000–400 cm^{-1} using potassium bromide. The optical absorption properties of the samples prepared in 350–800 nm range were collected on UV-vis DRS (SHIMADZU UV-3600 Plus). $BaSO_4$ is used as the reflectivity standard in UV-vis DRS experiment. Photoluminescence (PL) measurements of samples were carried out on SHIMADZU RF5301 (Japan), and data recorded in the range of 350–600 nm were recorded. The prepared samples were excited at 310 nm.

Keywords Hydrothermal method · Calcination method
Wet impregnation method

2.1 Preparation of Titanium-Based Photocatalysts

2.1.1 Preparation of V_2O_5/Titanium Dioxide Photocatalysts

V_2O_5/rutile-anatase photocatalyst system with different V/Ti element molar ratios ranging from 0:1 to 0.1:1 was prepared by using incipient wet impregnation method. In the typical synthesis (V/Ti element molar ratio is 0.02:1), ammonium partial vanadate (NH_4VO_3, 0.0585 g, 0.5 mmol) is added to 60 mL thermo deionized water and stirred for 15 min continuously. Then P_{25} titanium dioxide (1.9968 g, 25 mmol) was added to the solution for 30 min under continuous stirring and dried in 120 h at a drying oven. The dried samples were then ground with the agate mortar in the mortar, and the powder was calcined in muffle furnace. The muffle furnace temperature was 400 °C with the heating rate of 5 °C/min and held for 3 h. The 0.5, 1, 2, 4, 7, and 10 mol% of V-doped rutile-anatase nanocomposites hereafter were named as V-0.5, V-1, V-2, V-4, V-7 and V-10 respectively.

2.1.2 Preparation of Carbon Spheres Supported CuO/Titanium Dioxide Photocatalysts

Carbon spheres are prepared by improved hydrothermal synthesis. Typically, 0.1 M sucrose solution is filled into a 100 mL stainless steel autoclave with a filling rate of 90%. The autoclave was then put into the oven and maintained for 5 h at 180 degrees Celsius. After cooling in the air, the products were centrifugally separated and washed with Milli-Q water and absolute alcohol. Finally, they were dried at 80 °C for a whole night.

The synthesis process of CuO/titanium dioxide@C was as follows: Copper nitrate hemihydrate and glycerol (molar ratio 1:2) were added into ultrapure water (100 mL) to form hydrated copper (II)—glycerol complex. The amount of doping of CuO was set to 0.31 wt%. P25 titanium dioxide (5.0 g) and CSs were added to the solution under continuous stirring. The masses of CSs depended on the nominal CSs load. Then the copper glycerol complex was precipitated on the titanium dioxide carrier by adding 0.5 M NaOH under constant stirring until the pH value reached 12, and then added 2 mL ammonium hydroxide. The resulting suspension was stirred for one more hour, then the light gray powders (presumably $Cu(OH)_2$/titanium dioxide@C) were collected by centrifugal separation. After repeated washing with ultrapure water, the $Cu(OH)_2$/titanium dioxide@C powders were dried overnight at 70 °C in air. CuO/titanium dioxide@C photocatalysts were obtained by calcination of the $Cu(OH)_2$/titanium dioxide@C powders at 300 °C for 2 h.

2.1 Preparation of Titanium-Based Photocatalysts

2.1.3 Preparation of Carbon Decorated In_2O_3/Titanium Dioxide Photocatalysts

The modified photocatalyst with fixed carbon content (holding carbon precursor content fixed) and different In/Ti molar ratio ranging from 0:1 to 0.09:1 was prepared by a simple wet impregnation method. In a typical preparation process (In/Ti molar ratio was 0.02:1, i.e., 2 mol%), 20 mL distilled water and 30 mL anhydrous ethanol were mixed 5 min to obtain solution A in a continuous stirred beaker, and P25 titanium dioxide (25 mmol, 1.997 g) was added to the solution A in a continuous stirring. After that, glucose (0.5 g) and in $(NO3)$ $34.5H_2O$ (0.1910 g, 0.5 mmol) were mixed into the suspension B to obtain the suspension C. The suspension C was continuously stirred for 30 min and dried overnight in a 120 °C drying oven. After that, the samples were rubbed in the agate mortar with a pestle. The obtained powder was calcined in muffle furnace at a heating rate of 5 °C/min until it reaches 400 °C and maintains the temperature for 3 h. Finally, carbon modified In_2O_3/titanium dioxide nanocomposites were named C-In/Ti-2. We named 0, 1, 3, 6, and 9 mol% of In-doped titanium dioxide nanocomposites with carbon modification as C-P25, C-In/Ti-1, C-In/Ti-3, C-In/Ti-6, and C-In/Ti-9 respectively, and the sample C-In/Ti-2 without carbon doped as In/Ti-2. In_2O_3 was prepared by the same method.

2.2 Characterization of Titanium-Based Photocatalysts

2.2.1 Characterization of V_2O_5/Titanium Dioxide Photocatalysts

2.2.1.1 XRD Analysis

X-ray diffraction (XRD) patterns of various rutile-anatase homo-structure and V_2O_5/rutile-anatase samples are showed in Fig. 2.1. As the XRD pattern of P25 showed, diffraction peaks centered at $2\theta = 25.3°$, 37.8°, 48.0°, 53.9°, 55.1°, and 62.7° consistent with those from anatase titanium dioxide (101), (004), (200), (105), (211), and (204) planes respectively. While the peak centered at $2\theta = 27.4°$, 36.1°, 41.2°, 44.0° and 54.3° can be assigned to that from rutile phase (110), (101), (111), (210), and (211) planes for FP25 [1]. Diffraction peaks of both anatase titanium dioxide and rutile titanium dioxide are obtained, and anatase titanium dioxide is the dominant phase. Moreover, the mass ratio of rutile/anatase is almost no change when the calcination temperature is below 400 °C while that of rutile/anatase is increased when the calcination temperature is higher than 400 °C. In addition, no vanadium oxides (such as V_2O_5 and/or VO_2) diffraction peaks are identified as the V/Ti element molar ratios are less than 4 mol%, which means that either V is doped into rutile-anatase lattice or vanadium oxide is evenly dispersed. The uniformly dispersed active compo-

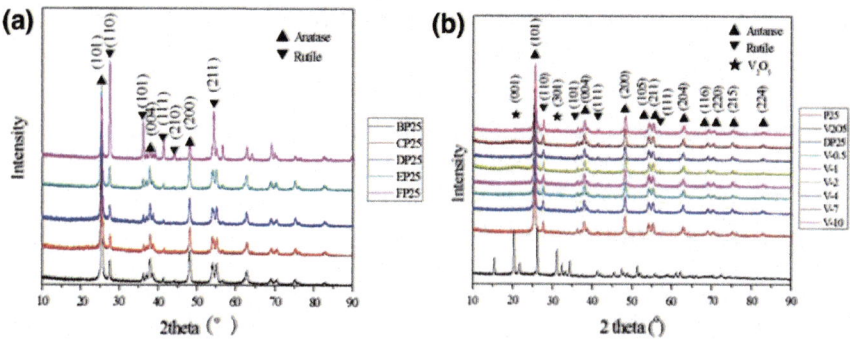

Fig. 2.1 XRD patterns of various samples with different **a** calcination temperature; **b** vanadium doping amounts. Reprinted from Ref. [2], Copyright 2017, with permission from Elsevier

nents in photocatalyst can enhance the photocatalytic activity of the host. However, when the doped vanadium oxides reach 7 mol% or more, weak diffraction peaks are centered at $2\theta = 20.3°$ and $31°$, which is consistent with V_2O_5 (001) and (301) planes. Ternary material of V_2O_5/rutile-anatase are confirmed and well match with the standard JCPDS data files No. 21-1272, 21-1276 and 41-1426 respectively.

2.2.1.2 The Appearance of as-Prepared Samples and SEM Analysis

The photograph of V_2O_5/rutile-anatase photocatalysts is described in Fig. 2.2a. At low V_2O_5 loading (0.5–2 mol%), the V_2O_5/rutile-anatase photocatalysts are all light yellow in color. When V_2O_5 loading exceeded 4 mol%, the V_2O_5/rutile-anatase photocatalysts are all yellow, and the intensity of yellow color increased proportionally with the V_2O_5 doping levels. In addition, the color of P25 is white, while the V_2O_5/rutile-anatase photocatalysts are yellow. It means that V_2O_5 has been successfully introduced into the titanium dioxide support and V_2O_5 nanoparticles has been formed on the surface of the titanium dioxide, which are confirmed by XRD data shown in Fig. 2.2.

The SEM images of V_2O_5/rutile-anatase photocatalysts with different V/Ti element molar ratios are shown in Fig. 2.2. There is no obvious difference in the morphology of all the V_2O_5/rutile-anatase samples. The SEM images of DP25 (Fig. 2.2b) show the hardened state of the spherical particles which are 15–40 nm in size. Compared to the image of DP25, the SEM image of V-2 depicted in Fig. 2.2c indicates that sphere-like particles are slightly dispersed. However, when the V/Ti molar ratios reach 0.07:1 or above, as shown in Fig. 2.2d, e, the shape and size of these particles are not modified compared with DP25, which mean that the V_2O_5/rutile-anatase photocatalysts are validated again. The SEM image of V_2O_5 (Fig. 2.2f) shows nanosheets with agglomerated status.

2.2 Characterization of Titanium-Based Photocatalysts

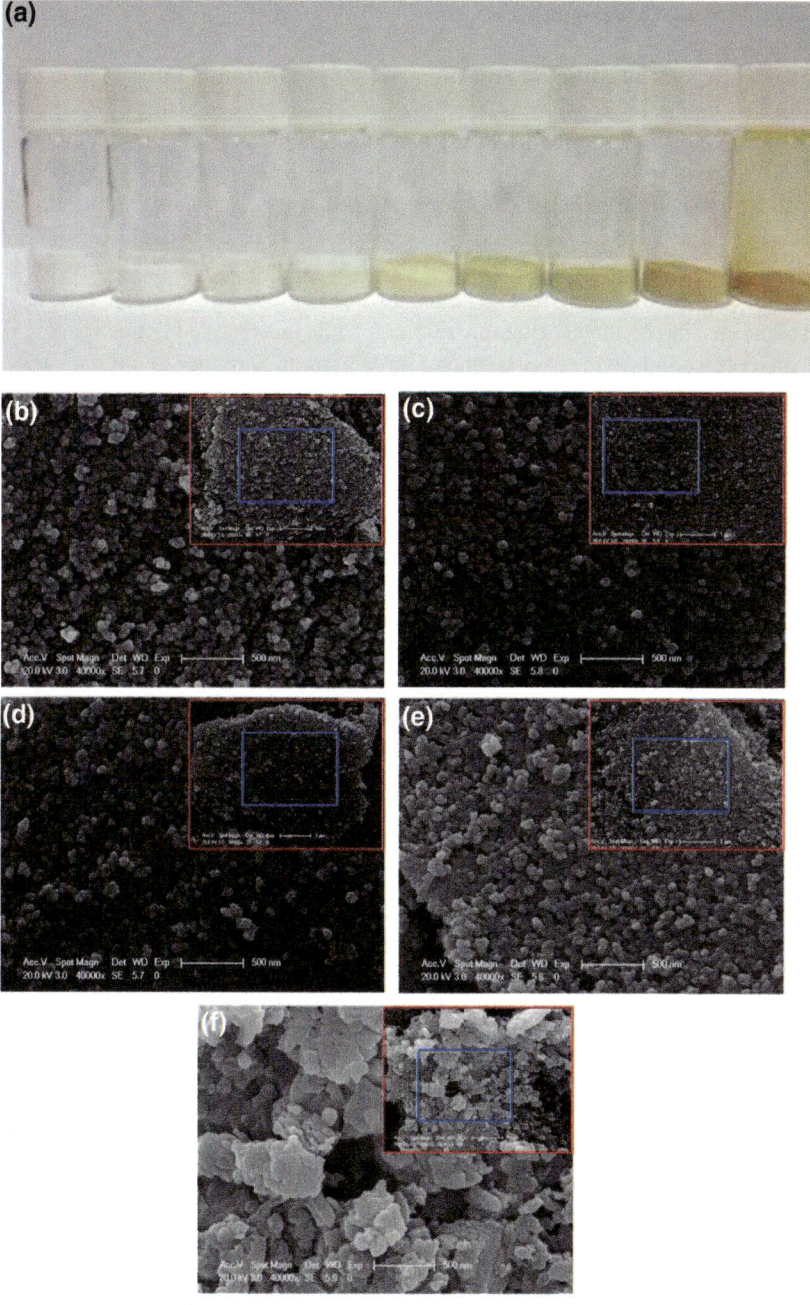

Fig. 2.2 **a** Photograph of the V_2O_5/rutile-anatase photocatalysts; SEM images of as-prepared samples: **b** DP25; **c** V-2; **d** V-7; **e** V-10; **f** V_2O_5. Reprinted from Ref. [2], Copyright 2017, with permission from Elsevier

2.2.1.3 TEM and EDS Analysis

To further investigate the microscopic morphology and structural information of the V_2O_5/rutile-anatase photocatalysts, transmission electron microscopy (TEM) analysis of V-2 and V-7 is performed, as shown in Fig. 2.3a, b. It can be seen that the V-2 and V-7 photocatalysts consist of nanoparticles with sizes between 15 and 40 nm, and V-2 sample are more dispersed compared with V-7, which agree well with the SEM images. The V_2O_5/rutile-anatase photocatalysts are characterized by HRTEM as well. Figure 2.3c, d show the magnified HRTEM images of V-2 and V-7 samples, which show clear lattice fringes. The lattice distance of 0.352 nm corresponds to the (101) plane spacing of anatase titanium dioxide, while the obtained lattice spacing of 0.325 nm corresponds to the (110) plane of rutile titanium dioxide. At the same time, V_2O_5 are not found on the surface of V-2 sample, indicating that the V_2O_5 are evenly dispersed on the titanium dioxide support, or even inserted into the lattice of titanium dioxide, which is in accordance with XRD pattern of V-2 sample. The XRD pattern of V-2 sample shows no peak of V_2O_5. It can be observed from Fig. 2.3d that the clear lattice fringes of V_2O_5 with lattice distance of 0.44 nm corresponds to the (001) plane. It means that V_2O_5 nanoparticles adhere to the surface of the titanium dioxide to form the heterostructure, which is consistent with the XRD results. To confirm the information about spatial distribution of Ti and V in the photocatalyst system, the elemental mapping analysis is used. As shown in Fig. 2.3e, f, the mappings of Ti and V elements are well defined and distributed uniformly throughout the nanocomposites, confirming that titanium dioxide and V_2O_5 have successfully combined to form the heterostructures. The energy dispersive x-ray spectroscopy (EDS) results shown in Fig. 2.3g, h indicate that the atomic ratio of Ti and V are 90.3 and 1.7% for V-2, 93.3 and 6.7% for V-7 respectively, which are same with theoretical atom ratio of Ti and V (Table 2.1).

2.2.1.4 BET Surface Areas Analysis

The structural characteristics of as-prepared V_2O_5/rutile-anatase nanocomposites were measured by N_2 adsorption-desorption isotherm. The results are shown in Fig. 2.4. All the isothermal are considered to be type IV hysteresis loop with high relative pressure range of 0.75–0.99, corresponding to mesoporous pore structures,

Table 2.1 The atomic percentages of Ti and V

Sample	V-2	V-7
Element	At%	At%
Ti K	90.3	93.3
V K	1.7	6.7
Totals	100	100

Reprinted from Ref. [2], Copyright 2017, with permission from Elsevier

2.2 Characterization of Titanium-Based Photocatalysts

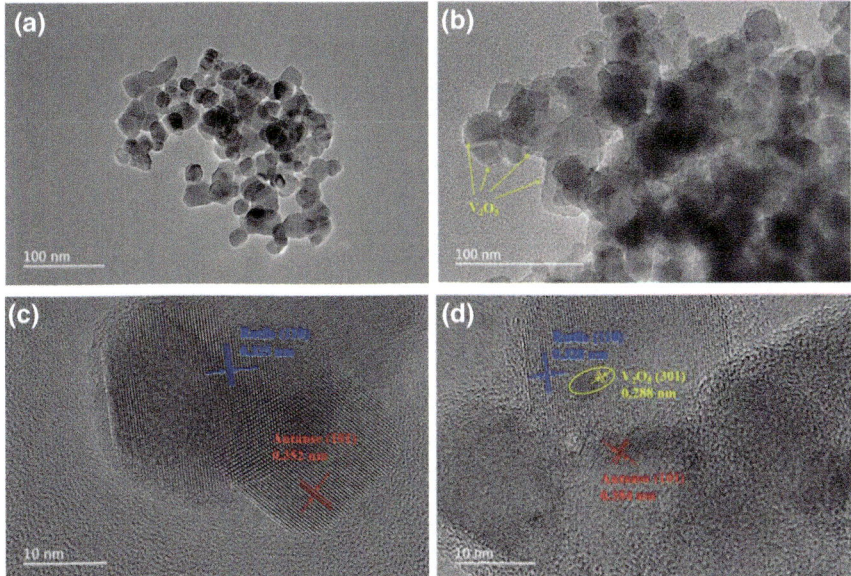

Fig. 2.3 TEM images of **a** V-2; **b** V-7; HRTEM images of **c** V-2; **d** V-7; **e** TEM and corresponding elemental mapping of V-2; **f** TEM and corresponding elemental mapping of V-7; EDS images of **g** V-2 and **h** V-7. Reprinted from Ref. [2], Copyright 2017, with permission from Elsevier

and the pore structure is mainly formed by the polymerization of titanium dioxide nanospheres, which is consistent with the SEM and TEM patterns [2]. The larger specific surface area usually contributes to the better photocatalytic reaction. The specific surface areas of V_2O_5/rutile-anatase nanocomposites are summarized in Table 2.2. The results showed that the specific area of the prepared V_2O_5/rutile-anatase was changed in the range of 12.63–47.26 m^2/g, and the specific surface areas of V-1 and V-2 are found to be the maximum value of 47.26 and 47.13 m^2/g. When the doping of vanadium is less than 1 mol%, the specific surface areas of the as-prepared samples universally increased with the increase of V_2O_5 loading. When the vanadium doping amount is above 2 mol%, the opposite phenomenon occurs. It may be that the excess vanadium species obstruct the stomata and active sites of the host, thus reducing the specific surface areas of the as-prepared V_2O_5/rutile-anatase photocatalysts.

2.2.1.5 UV-Vis Analysis

The ultraviolet-visible diffuse reflectance spectroscopy (UV-vis DRS) of pure titanium dioxide and V_2O_5-modified rutile-anatase samples are depicted in Fig. 2.5a. The results show that the absorption edges of all V_2O_5/rutile-anatase nanocomposites shift to the visible light range compared with that of pure titanium dioxide. At the same time, with the increase of active component V_2O_5, the absorp-

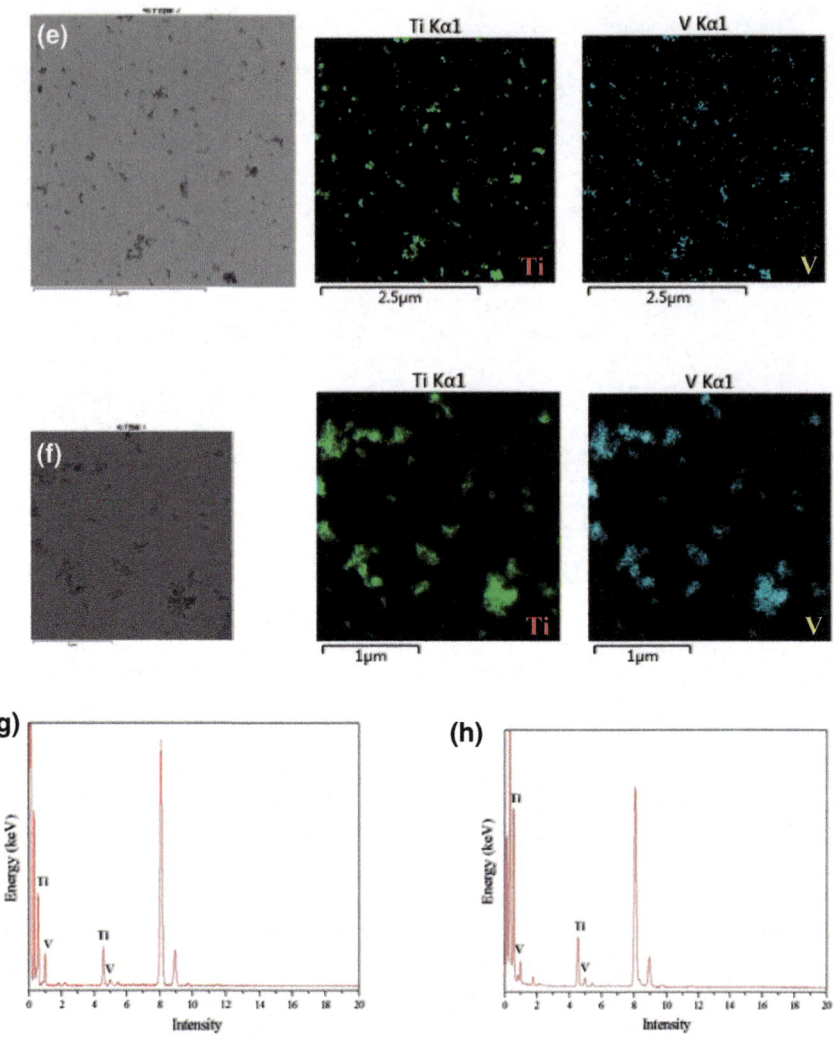

Fig. 2.3 (continued)

tion intensity also increased. Moreover, compared with pure titanium dioxide, V_2O_5/rutile-anatase nanocomposites have obvious light absorption ability in the range of UV and visible light. This is the reason why the oxygen vacancies in the preparation process may capture the additional electrons. These additional electrons have an important effect on the light absorption [3]. V_2O_5 modified titanium dioxide can form an impurity energy levels below conduction band of titanium dioxide [4]. As a result, it leads to forming a new conduction band at a lower energy, which narrows the band gap. The band gap energy (E_g) of the

2.2 Characterization of Titanium-Based Photocatalysts

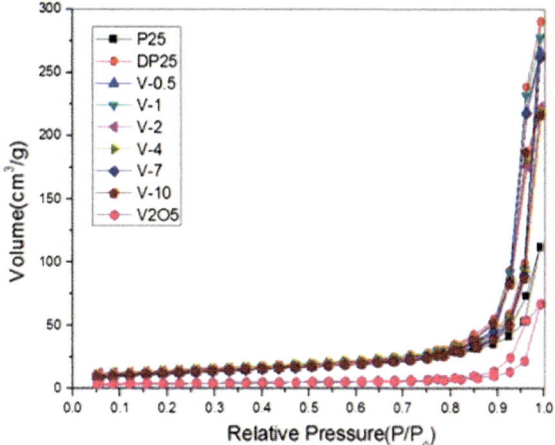

Fig. 2.4 N$_2$ adsorption-desorption isotherms of as-prepared V$_2$O$_5$/rutile-anatase photocatalysts. Reprinted from Ref. [2], Copyright 2017, with permission from Elsevier

Table 2.2 The physical, chemical and photocatalytic properties of the as-prepared V$_2$O$_5$/rutile-anatase nanocomposites

Samples	BET surface area (m^2/g)	Pore volume (cm^3/g)	Pore diameter (nm)	Absorbing boundary (nm)	Bandgap (eV)	Hg^0 oxidation efficiency (visible light) η (%)
P25	41.86	0.173	16.56	386	3.21	17.66
DP25	46.41	0.449	38.67	406	3.05	31.52
V-0.5	42.89	0.411	38.37	407	3.05	77.34
V-1	47.26	0.429	36.32	408	3.04	85.32
V-2	47.13	0.344	29.18	412	3.01	93.58
V-4	45.97	0.339	29.54	419	2.96	77.04
V-7	42.87	0.405	37.83	423	2.93	23.88
V-10	43.39	0.334	30.76	442	2.81	14.50
V$_2$O$_5$	12.63	0.104	32.86	–	2.14	9.03

Reprinted from Ref. [2], Copyright 2017, with permission from Elsevier

V$_2$O$_5$/rutile-anatase photocatalysts can be computed according to Kubelka-Munk equation [5]:

$$E_g = \frac{1240}{\lambda_{Absorp.Edge}} \quad (2.1)$$

The band gap energy values are displayed in Table 2.3. The data show that the absorption edge of DP25 prepared by the same method without vanadium precursor shift to lower energy compared with P25. The reason is that the homojunction is formed between anatase titanium dioxide and rutile titanium dioxide during the

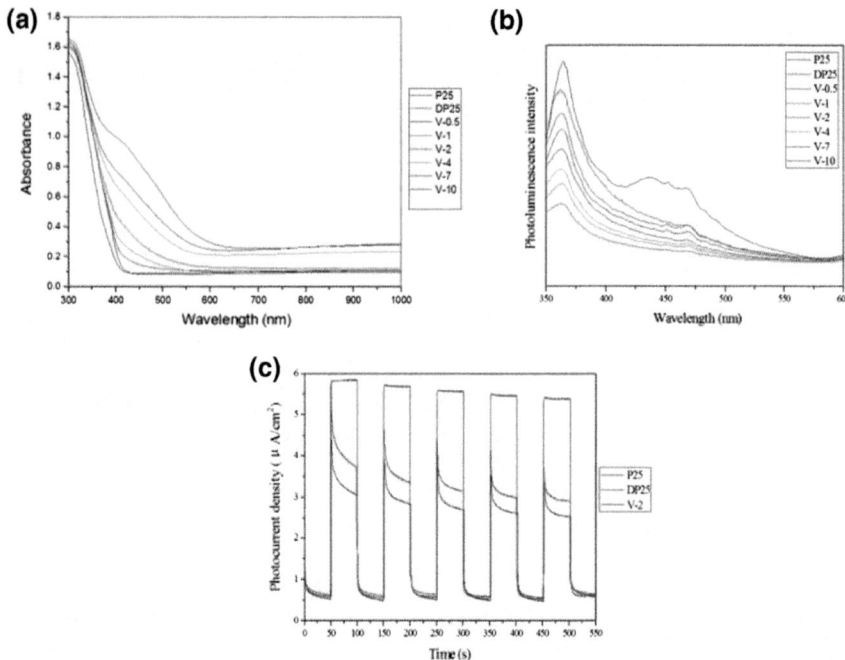

Fig. 2.5 **a** UV-vis absorbance spectra; **b** Photoluminescence emission spectra; **c** Transient photocurrent response of as-prepared V_2O_5/rutile-anatase photocatalysts under visible light irradiation ([Na_2SO_4] = 0.5 M). Reprinted from Ref. [2], Copyright 2017, with permission from Elsevier

Table 2.3 FTIR spectral assignments of as-prepared nanocomposites

Band position (cm^{-1})	Spectral assignment
3600–3200	O-H stretching of physically adsorbed water molecule
1630	H-O-H bending of physically adsorbed water molecule
1400	Ti-O-Ti stretching vibration
1020	V=O stretching vibration
830	V-O-V stretching vibration
675–505	Ti-O-Ti stretching vibration

Reprinted from Ref. [2], Copyright 2017, with permission from Elsevier

preparation process, which is confirmed by the HRTEM pattern shown in Fig. 2.3c. At the same time, the absorption edge of the V_2O_5/rutile-anatase photocatalysts progressively shifts to lower energy with the increasing levels of V_2O_5.

2.2.1.6 Optical and Photoelectrochemical Properties

In general, the photocatalytic activity of the samples greatly depends on the separation efficiency of the photogenerated electrons and holes. Photoluminescence (PL) emission spectrum is adopted to research the efficiency of separation capability of excited electrons and holes. Lower PL emission intensity means higher excited electrons and holes separation ability, and the separation of electrons and holes extremely improves the photocatalytic activity of photocatalysts [6]. PL emission spectra of the as-prepared V_2O_5/rutile-anatasesamples are shown in Fig. 2.5b. As can be seen in Fig. 2.5b, the intense of PL emission is surveyed for the pure P25, indicating that photogenerated electron-hole pairs recombinated rapidly. Whereas, the peak intensities of the as-prepared V_2O_5/rutile-anatase photocatalysts are significantly weaker than pure P25, which indicates that charge carriers are of more efficient transfer and separation due to the fact that the excited electrons and holes transfer in the homo-hetero junctions among V_2O_5, rutile titanium dioxide, and anatase titanium dioxide, meaning that V_2O_5 effectively suppressed electron-hole pairs recombination. In order to further research the separation and migration of photoinduced electrons and holes, photocurrent responses are carried out. As shown in Fig. 2.5c, both DP25 and V-2 exhibit enhanced photocurrent responses than the P25. Therefore, the separation of photoexcited electrons and holes is improved for DP25 by the interfacial interactions [7, 8]. Furthermore, the photocatalyst system can further enhance the separation of electron-hole pairs than two-component systems. The separated holes combining with OH^- produce •OH radicals, which can enhance the photocatalytic activity of as-prepared samples. Further inspection of PL spectra reveal that V-2 sample have the lowest PL intensity among the as-prepared samples, which indicates that the separation of excited electrons and holes are promoted highest.

2.2.1.7 Surface Functional Groups of the as-Prepared Samples

Fourier transform infrared spectroscopy (FTIR) spectra of nanocomposites are shown in Fig. 2.6. The corresponding spectral distributes for various peaks surveyed are given in Table 2.3. The sharp peak at 1630 cm^{-1} and a broadband in area between 3600 and 3200 cm^{-1} are caused by the stretching vibration of the H-O-H bending and O-H of physical adsorption water respectively. The peak at 1400 cm^{-1} and a broadband in area between 675 and 505 cm^{-1} are attributed to the Ti-O-Ti stretching vibrations [9]. Pure vanadia spectra show stretching mode of $V=O$ at 1020 cm^{-1} and the V-O-V stretching mode at 830 cm^{-1} [10]. Except for the pure vanadia and V-10 samples, the peak value appeared at 1020 cm^{-1}, which vanished in V-2 sample. This observation is excellently same with the XRD data.

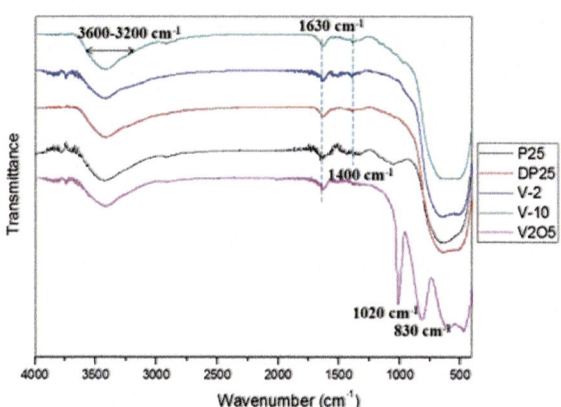

Fig. 2.6 FTIR spectra of as-prepared V$_2$O$_5$/rutile-anatase photocatalysts. Reprinted from Ref. [2], Copyright 2017, with permission from Elsevier

2.2.1.8 X-Ray Photoelectron Spectroscopy (XPS) Analysis

The surface composition and chemical status of the as-prepared V$_2$O$_5$/rutile-anatase photocatalysts are investigated by XPS analysis. The high resolution spectrum of O 1s for V-2 and V-10 are presented in Fig. 2.7a, b. The peaks can be fitted into two different oxygen species. The main peak is assigned to lattice oxygen for titanium dioxide. Another peak is indicated to the surface-adsorbed hydroxide (OH$^-$) on the titanium dioxide surface [11]. These surface-adsorbed hydroxide are also confirmed by FTIR. However, the energy peaks of lattice oxygen in V-2 and V-10 are all higher than that of lattice oxygen in titanium dioxide. It can be attributed to oxygen density decreasing resulted from the doped V ions substituting Ti ions while V element has higher electronegativity. In view of the above analysis, it is further proof that the V ions have adulterated into titanium dioxide lattice to replace Ti ions, which is of accordance with XRD data. It is well known that the OH$^-$ play an important role in photocatalytic reaction. The reason is that it can engender •OH by catching the excited holes, which benefits to the oxidation reaction [12]. Therefore, in the process of preparing nanocomposites, the physical absorption capacity of the hydroxyl group not only enhances the splitting efficiency of the electron hole pair, but also promotes the transfer of the excited hole to the hydroxyl group, thus improving the efficiency of the photocatalytic reaction. As shown in Fig. 2.7c, d, compared with Ti^{4+} ions in titanium dioxide support, the binding energies of the Ti 2p region shift to a subtractive value, which indicates that V element incorporates into the lattice [13]. The XPS spectrum of the V 2p region for V-2 and V-10 are shown in Fig. 2.7e, f. The peak located at about 515.44 eV for V-2 and 516.25 eV for V-10 are attributed to V^{4+} of the V 2p$_{3/2}$ electronic state, the peaks at about 516.43 eV for V-2 and 517.23 eV for V-10 can be assigned to V^{5+} 2p$_{3/2}$, and the peaks at about 521.19 eV for V-2 and 521.37 eV for V-10 are attributed to V^{5+} of the V 2p$_{1/2}$ [14]. Obviously, the ratio of V^{5+}/V^{4+} for V-2 is higher than the V-10, which indicates that V-2 possessed more quantity of higher oxidation state of the element V. According to Mars and Van Krevelen mechanism, higher oxidation state of element V can lead to better photocatalytic

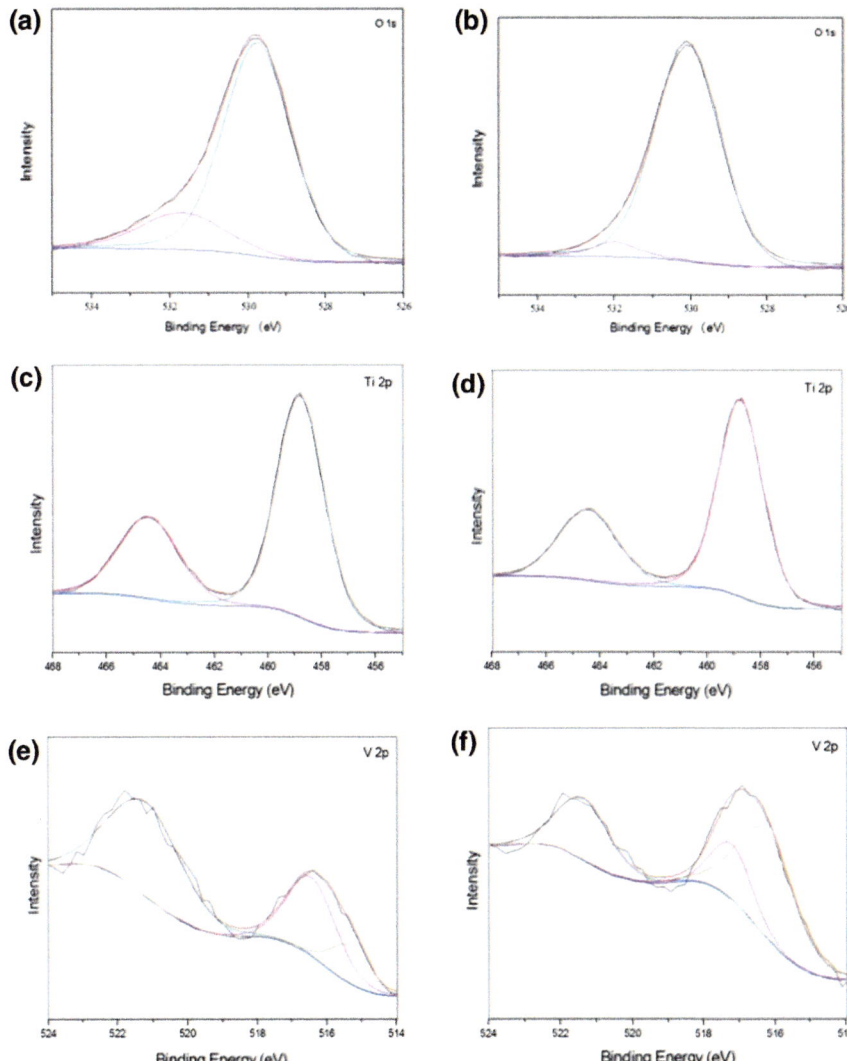

Fig. 2.7 a O 1s XPS spectra of fresh V-2; b O 1s XPS spectra of fresh V-10; c Ti 2p XPS spectra of fresh V-2; d Ti 2p XPS spectra of fresh V-10; e V 2p XPS spectra of fresh V-2; f V 2p XPS spectra of fresh V-10. Reprinted from Ref. [2], Copyright 2017, with permission from Elsevier

activity [15], which means that V-2 has a higher photocatalytic oxidation efficiency than V-10.

2.2.2 Characterization of Carbon Spheres Supported CuO/Titanium Dioxide Photocatalysts

2.2.2.1 Formation Mechanism and SEM Observation

The preparation of CuO/titanium dioxide@C photocatalyst is shown in Fig. 2.8. A dense layer of titanium dioxide particles was directly deposited on the carbonspheres (CSs) with the ammonia ions (NH_4^+). The P25 coating formation mechanism includes the following steps: (1) Carbonspheres (CSs) were prepared by sucrose via hydrothermal synthesis. (2) In the subsequent coating process, CSs was dispersed into $Cu(OH)_2$/titanium dioxide solution and formed $Cu(OH)_2$/titanium dioxide@C by NH_4^+. (3) Finally, CuO/titanium dioxide@C was produced by calcination.

To attain C content in the CuO/titanium dioxide@C photocatalysts and the nanoporous structures, the Fig. 2.9 shows the FESEM/EDAX micrographs of CuO/titanium dioxide@C. Figure 2.9a shows an image of CSs. As can be seen, large quantities of CSs were produced, with an average diameter of ca. 700–900 nm. Compared with the P25, there was no significant difference in the morphology of catalyst with different ratio of CSs doping. When the concentration of doped CSs was 0.25%, no CSs were found. The reason may be that the content is too low to detect the core nanocomposites. As shown in Fig. 2.9, Cu, C, Ti, and O are present in all samples. As the doping of CSs increases, the C content in the sample also gradually increases. Unfortunately, EDAX cannot accurately detect the concentration of doped CSs because the CSs are much larger than titanium dioxide and the distribution of CSs is not uniform.

2.2.2.2 XRD Analysis

Figure 2.10 shows the XRD patterns of the as-prepared CuO/titanium dioxide@C samples. All samples are composed of anatase and rutile. When the CuO load-

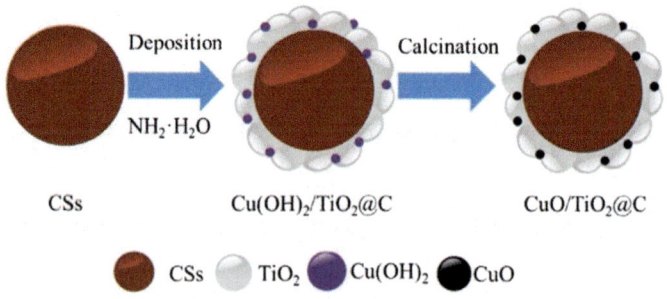

Fig. 2.8 Illustration of synthetic procedure for the CuO/ titanium dioxide@C. Reprinted from Ref. [16], Copyright 2016, with permission from Elsevier

2.2 Characterization of Titanium-Based Photocatalysts

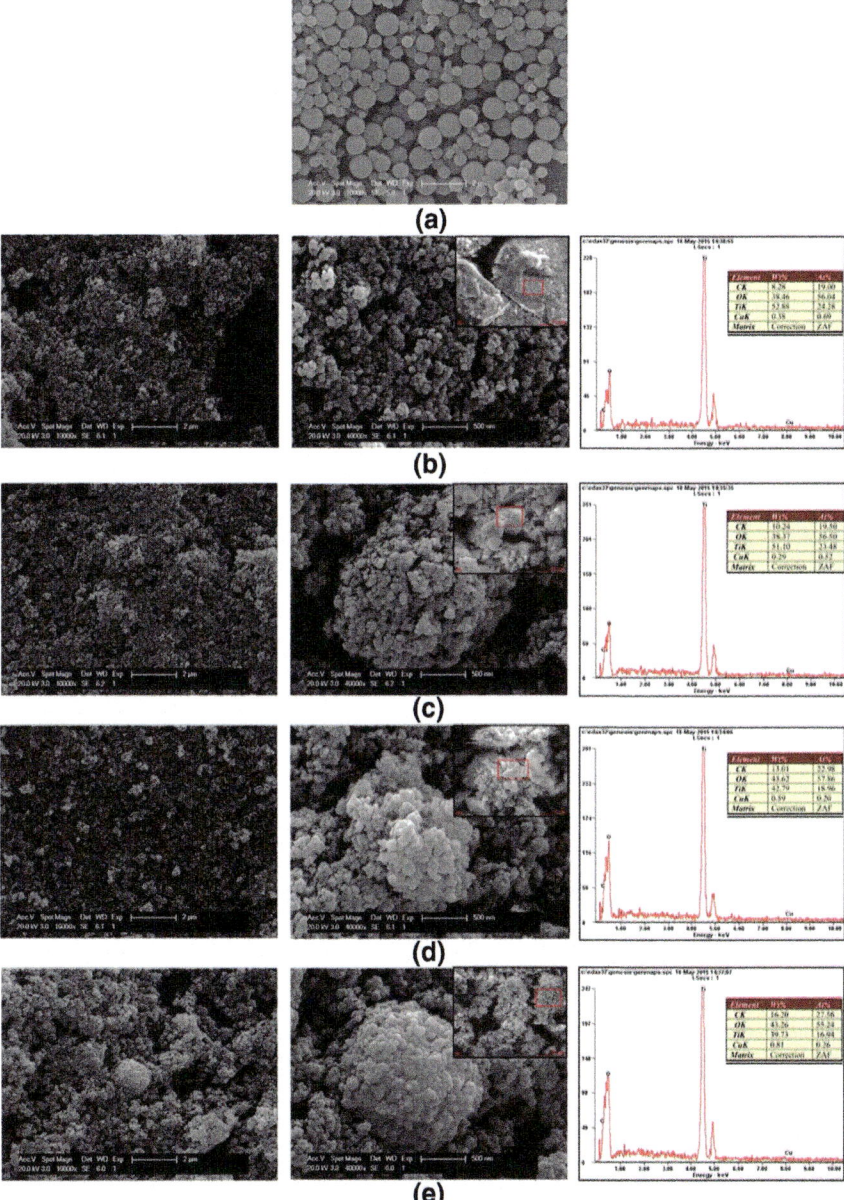

Fig. 2.9 FESEM images (left) and EDAX spectra (right) for selected CuO/titanium dioxide@C photocatalysts: **a** CSs; **b** CuO/titanium dioxide@0.25 wt.%C; **c** CuO/titanium dioxide@0.5 wt.%C; **d** CuO/titanium dioxide@1 wt.%C; **e** CuO/titanium dioxide@2 wt.%C. Reprinted from Ref. [16], Copyright 2016, with permission from Elsevier

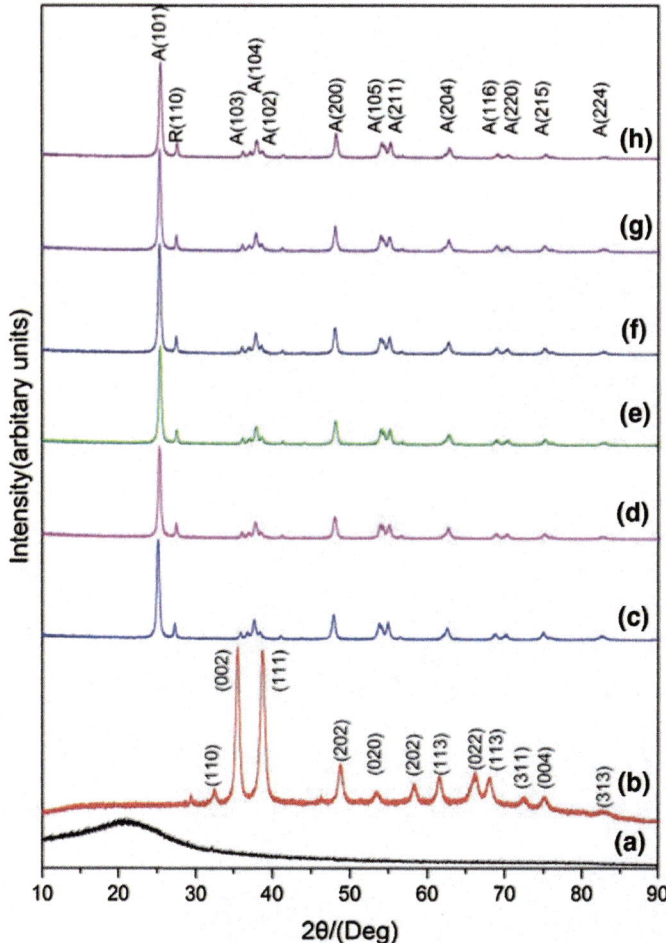

Fig. 2.10 Powder XRD patterns for the 0–2 wt.% CuO/titanium dioxide@C photocatalysts after calcination at 300 °C for 2 h: **a** CSs; **b** CuO; **c** P25 titanium dioxide; **d** CuO/titanium dioxide; **e** CuO/titanium dioxide@0.25 wt.%C; **f** CuO/titanium dioxide@0.5 wt.%C; **g** CuO/titanium dioxide@1 wt.%C; **h** CuO/titanium dioxide@2 wt.%C. Reprinted from Ref. [16], Copyright 2016, with permission from Elsevier

ing concentration is low, no diffraction peak corresponding to CuO is found in the CuO/titanium dioxide@C composite photocatalyst (0.31 wt.%) [16]. The XRD patterns of all these samples provide similar profiles, indicating that the introduction of CSs and CuO loading have no significant effect on the structure of titanium dioxide. The average crystal size of all samples was estimated based on the Scherrer formula as follows:

$$D = \frac{k\gamma}{\rho \cos \theta} \qquad (2.2)$$

where D is crystalline size, K=0.89 is a coefficient, γ is the X-ray wavelength corresponding to the Cu Kα radiation, ρ is the half-height width of the diffraction peak of anatase or rutile respectively, 2θ=25.3. Based on the Prague equation and the use of XRD software, the unit cell parameters of all samples have been analyzed. As a result, the crystallize sizes of samples calculated via the Scherrer formula were 212, 223, 223, 217, 228, and 219 Å, respectively represent P25, CuO/titanium dioxide, CuO/titanium dioxide@ywt.%C (y=0.25; 0.5; 1; 2). It seems that with CSs doping, the crystallize sizes become small [17]. This shows that CSs can be used as a dispersant to inhibit the growth of grain, and to load less crystalline CuO/titanium dioxide domain on the surface of CSs, thus avoiding the aggregation of CuO/titanium dioxide.

Figure 2.11 showed an expanded view of the XRD patterns in the 2θ range 24–29.5. It can be seen from the diagram that in some cases, the corresponding intensities and positions of several diffraction peaks in samples are different. The position of A (101) and R(110) reflection in CuO/titanium dioxide@C shows a little shift to higher angle, which may be ascribed to the lattice expansion caused by a tiny fraction of the carbon atoms incorporation. It indicates that in the reaction process, the CSs have a certain influence on the A(101) and R(110) space of titanium dioxide.

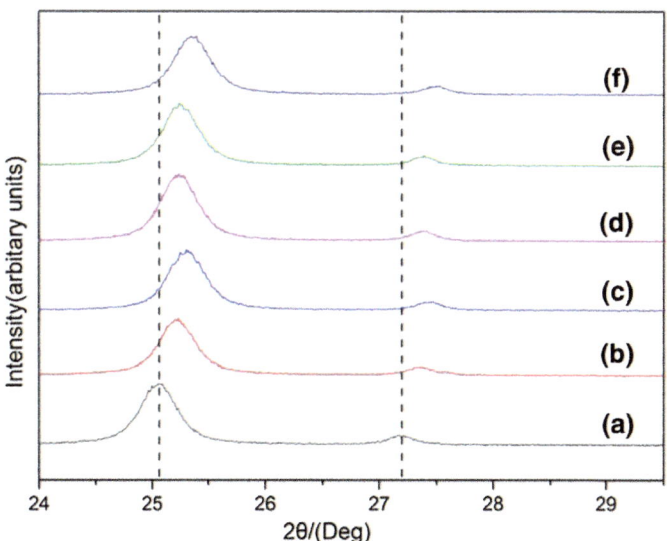

Fig. 2.11 Powder XRD patterns for the 0–2 wt.% CuO/titanium dioxide@C photocatalysts after calcination at 300 °C for 2 h: **a** P25 titanium dioxide; **b** CuO/titanium dioxide; **c** CuO/titanium dioxide@0.25 wt.%C; **d** CuO/titanium dioxide@0.5 wt.%C; **e** CuO/titanium dioxide@1 wt.%C; **f** CuO/titanium dioxide@2 wt.%C. Reprinted from Ref. [16], Copyright 2016, with permission from Elsevier

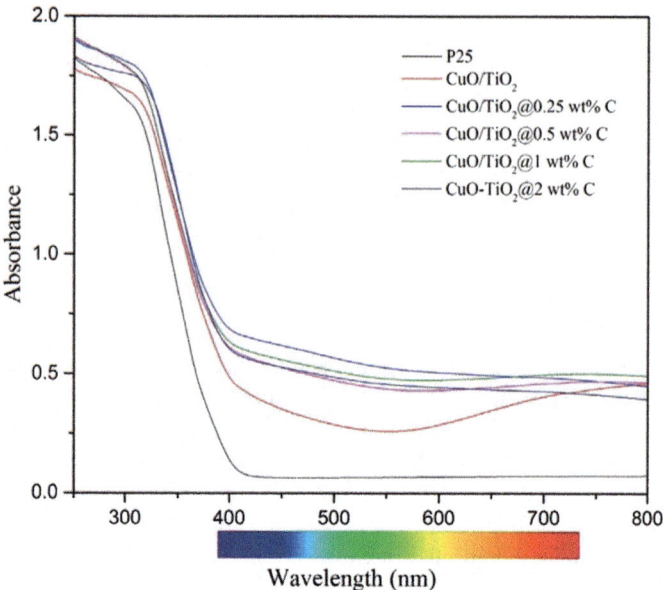

Fig. 2.12 UV-vis absorbance spectra of prepared CuO/titanium dioxide@C samples. Reprinted from Ref. [16], Copyright 2016, with permission from Elsevier

2.2.2.3 UV-Vis DRS and BET

Using the spectrum of P25 as reference material, the electronic state is further measured by ultraviolet visible spectroscopy. Figure 2.12 has indicated that there is not any adsorption for P25 over their fundamental absorption band edge 410 nm. Significant red-shift in the UV-vis absorption spectra is observed for CuO doped titanium dioxide. It is well known that the absorption properties of carbon elements have been known. Due to the doping of carbon spheres, more red shifts appear in the UV absorption spectra in the synthesized core-shell nanocomposites [18]. Similar red-shift for carbon-doped titanium dioxide has been also confirmed by previous investigators [18, 19]. Lastly, the optical band gap energy (E_g) of the CuO/titanium dioxide@C photocatalysts were calculated based on the absorption spectrum of the samples according to equation of $E_g = 1240/\text{Absorp.Edge}$ [5]. The results were shown in Table 2.4 The data showed that the absorption edge of the CuO/titanium dioxide@C photocatalysts progressively shifted to lower energy with increasing levels of CSs loading.

Figure 2.13 demonstrates that the N_2 adsorption/desorption isotherms of CuO/titanium dioxide@C photocatalysts. Table 2.4 summarized N_2 physisorption data for the CuO/titanium dioxide@C photocatalysts. All samples, including the titanium dioxide, had similar N_2 physisorption isotherms that could be classified as Type II accordingly to IUPAC convention for adsorption isotherms. When the CuO loading increases, the BET surface area of the photo catalyst usually increases. How-

2.2 Characterization of Titanium-Based Photocatalysts

Table 2.4 Summary of the physical, chemical and photocatalytic properties of the CuO/titanium dioxide@C photocatalysts

CSs loading (wt.%)	Absorbing boundary (nm)	Bandgap (eV)	C content by EDAX (wt.%)	BET surface area (m²/g)	Pore volume (cm³/g)	Pore diameter (nm)	Hg^0 removal efficiency (visible light) η (%)
TiO_2	394	3.15	–	47.65	0.176	16.02	19.9
CuO/TiO_2	435	2.84	–	43.83	0.460	43.07	54
0.25	443	2.81	8.28	43.59	0.384	36.77	62.0
0.5	448	2.77	10.24	48.23	0.434	36.47	64.1
1	451	2.75	13.01	50.09	0.429	36.96	61.3
2	455	2.73	16.20	47.84	0.417	39.04	44

Reprinted from Ref. [16], Copyright 2016, with permission from Elsevier

ever, when the doping amount is 2, the surface area of the photo catalyst is lower than that of the others. It can be explained that when the CSs doping was 2 wt.%, the Hg^0 removal efficiency decreased. This viewpoint will be elaborated in detail below.

2.2.2.4 FT-IR Spectroscopy

The bonding characteristics of functional groups in undoped titanium dioxide and CuO/titanium dioxide@C are identified by FT-IR spectroscopy (Fig. 2.14). The absorption peaks at about 3400 and 1630 cm^{-1} are associated with the stretching vibrations of surface water molecules, including hydroxyl groups and molecular water on the samples. Further observation shows that the peaks corresponding to the stretching vibration of water and hydroxyl groups are broader and stronger in the CuO/titanium dioxide@C than that of CuO/titanium dioxide and P25. The surface hydroxyl groups play an important role in the photocatalytic process, as these groups can be trapped by the holes generated under irradiation to form hydroxyl radicals which can suppress electron-hole recombination, increasing the photocatalytic efficiency [20].

2.2.2.5 XPS Analysis

The presence of carbon in CuO/titanium dioxide nanoparticles was confirmed by elemental analysis (EA) and X ray photoelectron spectroscopy (XPS). Figure 2.15a, can clearly identify the existence of C, Ti and O elements with the combination of C 1s, Ti 2p and O 1s, indicating that the prepared CuO/titanium dioxide @C contains carbon. As shown in Fig. 2.15b, the peaks at 464.93 and 458.85 eV in the Ti 2p high-resolution XPS spectra correspond to Ti 2p1/2 and Ti 2P3/2 respectively.

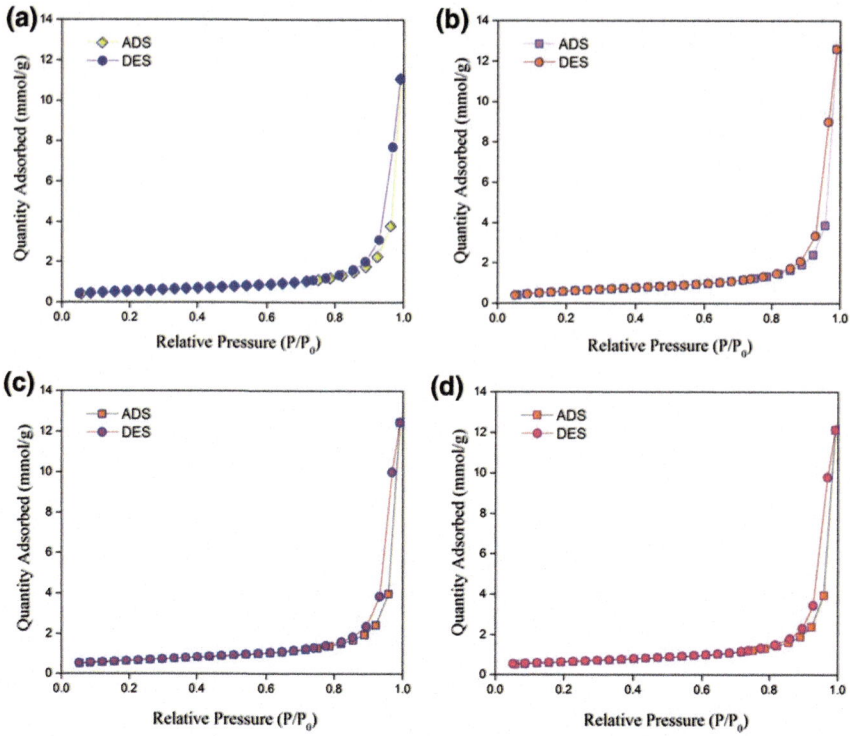

Fig. 2.13 N$_2$ adsorption/desorption isotherms of CuO/titanium dioxide@C photocatalysts prepared with different amount of CSs added. **a** 0.25 wt.%C; **b** 0.5 wt.%C; **c** 1 wt.%C; **d** 2 wt.%C. Reprinted from Ref. [16], Copyright 2016, with permission from Elsevier

The split of Ti 2p1/2 and Ti 2P3/2 is 6.08 eV, indicating the state of Ti^{4+} in the CuO/titanium dioxide doped with carbon spheres [21]. In addition, compared with the undoped P25 (Degussa) sample (464.0 eV) in literature [19], the binding of Ti 2p1/2 can increase 0.93 eV. The binding energies of CuO/titanium dioxide @ C increase, strongly indicating lattice distortion [19]. Figure 2.15c shows that the main C 1s peak is dominated by the element carbon of 284.43 eV, being mainly attributed to the alternating hydrocarbon in the extensive delocalization, and the two peaks at the 286.27 and 288.50 eV are the characteristics of the oxygen bound CO and Ti-OC, respectively [18, 19]. Thus, there are many kinds of carbon species in the lattice of CuO/titanium dioxide@ C, that is, substitution and interstitial carbon atoms and carbonate species [22]. On the other hand, the O 1s binding energy (Fig. 2.15d) of C doped samples also showed a marked change compared with the O 1s binding energy of undoped P25 (Degussa). This is from 529.6 eV obtained from a previous products in the undoped titanium dioxide materials to 529.99 eV in the C-doped hollow titanium dioxide sample and from 530.9 eV in the undoped to 531.78 eV in the doped sample. It shows that lattice distortion in doping is existed once more [23].

2.2 Characterization of Titanium-Based Photocatalysts 33

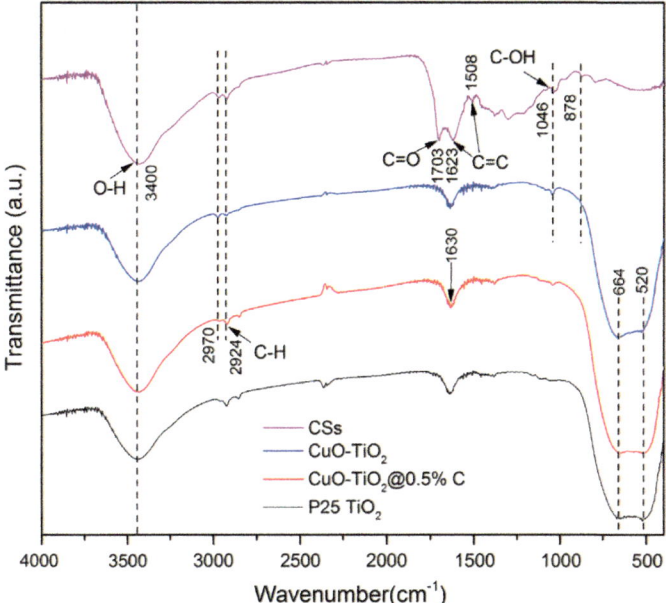

Fig. 2.14 FT-IR spectra of CSs, P25 titanium dioxide, CuO-titanium dioxide and CuO-titanium dioxide@0.5 wt.%C. Reprinted from Ref. [16], Copyright 2016, with permission from Elsevier

2.2.2.6 Photoluminescence Spectra

Photoluminescence (PL) emission spectroscopy was widely used to study carrier transfer efficiency and lifetime of photoelectron holes in semiconductors. Diverse studies using the PL spectrum also demonstrated a significant improvement in the photoinduced carrier separation (as shown in Fig. 2.16) [24]. It can be seen from the graph that the highest peak represents commercial P25, which indicates that the electrons and holes produced on P25 are easy to compound. After doping CuO with titanium dioxide, the electron and hole recombination of the catalyst decreased. With the further increase of the amount of carbon sphere, the recombination of electrons and holes of the catalyst decreased, and the photocatalytic activity of the photocatalyst was improved under the visible light photocatalytic performance.

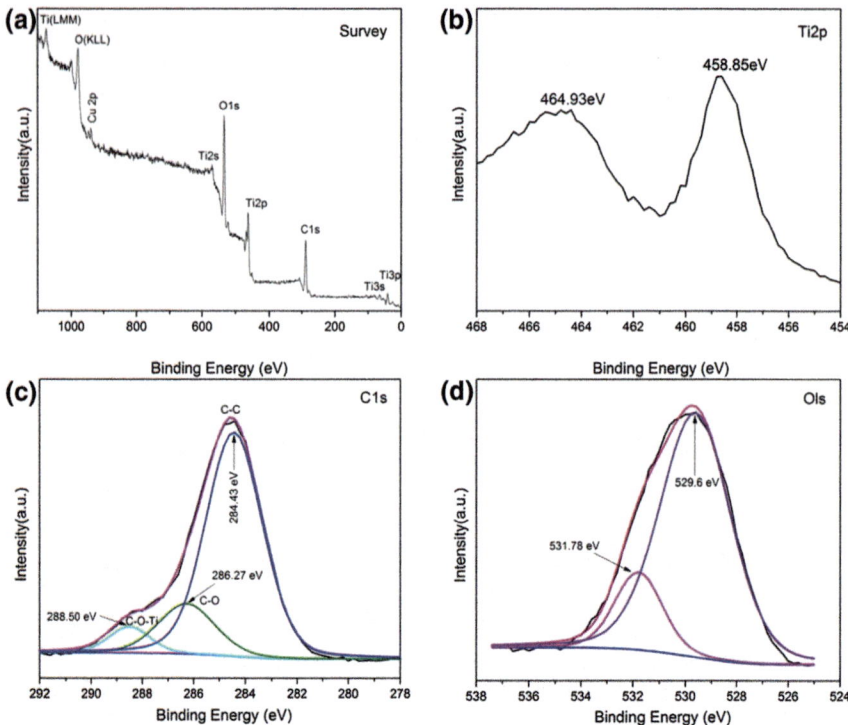

Fig. 2.15 a XPS fully scanned spectra of the CuO/titanium dioxide@0.5 wt.%C; **b** XPS spectra of Ti 2p; **c** XPS spectrum of C 1s and **d** XPS spectra of O 1s for the CuO/titanium dioxide@0.1 wt.%C samples. Reprinted from Ref. [16], Copyright 2016, with permission from Elsevier

2.2.3 Characterization of Carbon Decorated In_2O_3/Titanium Dioxide Photocatalysts

2.2.3.1 Crystalline Phase Analysis

In order to confirm the chemical composition of the prepared heterostructures, the C-In/Ti-2 of the studied samples were further studied. The surface chemical composition and chemical state of C-In/Ti-2 samples were characterized by X ray photoelectron spectroscopy (XPS) analysis. As can be seen from Fig. 2.18a, the measured spectra confirm that In, Ti, O and C elements exist in C-In/Ti-2 samples. Figure 2.18b, e is a high-resolution spectrum of C 1s, O 1s, Ti 2p and In 3D of nanocomposites. High resolution C 1s spectra can be observed in Fig. 2.18b, where the centers of the three peaks are at 288.0285.5 and 284.6 eV respectively. The peak value at 284.6 eV is set to the reference value of adjustment, and the peak at 285.5 eV corresponds to oxygen binding material C-O [25]. Therefore, the peak value of doped carbon is 288.0 eV, which replaces the main oxygen ions and forms the O-Ti-C bond [26].

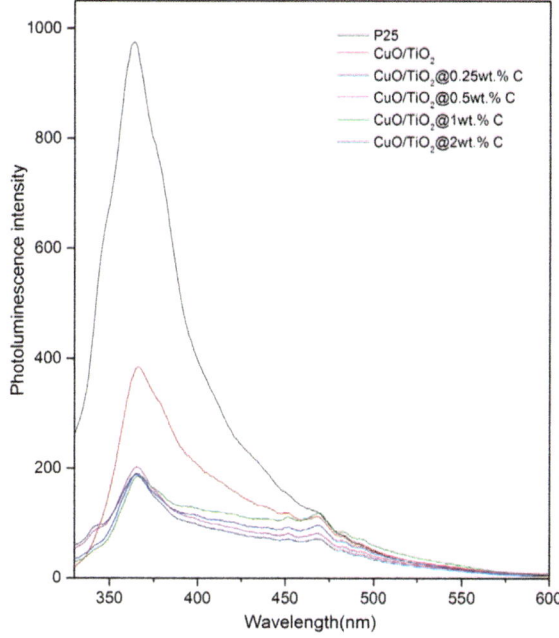

Fig. 2.16 UV-vis absorbance spectra for the CuO/titanium dioxide@C photocatalysts. Reprinted from Ref. [16], Copyright 2016, with permission from Elsevier

Figure 2.18c gives the O 1s photoelectron peak, in which there are four oxygen signals. The O 1s peak at 529.5 eV can correspond to the lattice oxygen in titanium dioxide, and the peak at 530.5 eV can point to the oxygen anion from In_2O_3, and the peak of the 531.4 eV is the characteristics of the oxygen binding species CO, and the peak of the surface hydroxyl is at 532.1 eV [27, 28]. In sample C-In/Ti- 2, the peak at 463.7 eV is associated with Ti 2p1/2, and another peak at 458.0 eV corresponds to Ti 2P3/2, and its spacing is 5.7 eV, indicating that standardization forms a state of Ti^{4+} in the In_2O_3/titanium dioxide heterostructure. The In 3D peaks at 444.2 and 451.7 eV represent In3d5/2 and In3d3/2, and correspond to In_3^+ in In_2O_3. Finally, it is confirmed that the three element materials of C, titanium dioxide and In_2O_3 are in the preparation of nanocomposites (Fig. 2.17).

2.2.3.2 XPS Analysis

To confirm the chemical composition of the as-prepared heterostructure, the sample C-In/Ti-2 is further studied. The X-ray photoelectron spectroscopic (XPS) analysis is adopted to characterize the surface chemical compositions and chemical states of the C-In/Ti-2 sample. As observed from Fig. 2.18a, the survey spectra confirm In, Ti, O, and C elements are existed in C-In/Ti-2 sample. Figure 2.18b, e provide the high resolution spectra of C 1s, O 1s, Ti 2p and In 3d for the nanocomposite. High resolution C 1s spectrum can be observed in Fig. 2.18b, in which the centers of the

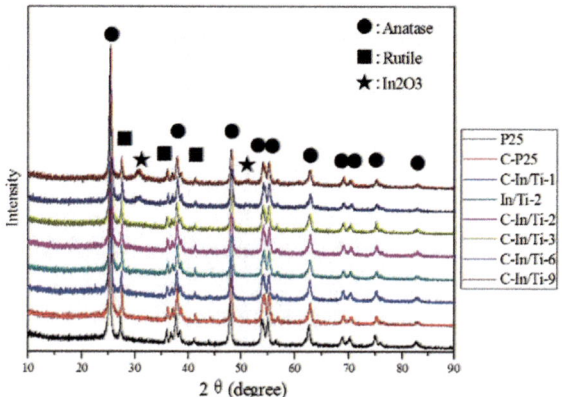

Fig. 2.17 XRD patterns of as-prepared carbon modified In_2O_3/titanium dioxide nanocomposites. Reprinted from Ref. [29], Copyright 2017, with permission from Elsevier

three peaks are at 288.0, 285.5, and 284.6 eV respectively. The peak at 284.6 eV is set as the adjusted reference and the peak at 285.5 eV is corresponding to the oxygen bound species C-O [25]. Accordingly, the peak at 288.0 eV is assigned to the doped carbon, which substituted the oxygen ion of the host, forming the O-Ti-C bond [26]. Figure 2.18c presents the O 1s photoelectron peaks, in which there exist four oxygen signals. The O 1s peak at 529.5 eV can be corresponding to lattice oxygen in titanium dioxide, the peak at 530.5 eV can be indexed to oxygen anions from In_2O_3, the peak at 531.4 eV can be associated to oxygen bound species C-O, and the peak at 532.1 eV is the characteristics of the surface hydroxyl groups [27, 28]. In the sample C-In/Ti-2, the peak at 463.7 eV is associated to the Ti $2p_{1/2}$ and the other peak at 458.0 eV is corresponding to Ti $2p_{3/2}$, of which spacing is 5.7 eV, indicating that the standardized state of Ti^{4+} forms in the In_2O_3/titanium dioxide heterostructure [30]. The In 3d peaks at 444.2 and 451.7 eV represent $In3d_{5/2}$ and $In3d_{3/2}$ and correspond to In^{3+} in In_2O_3. Lastly, the ternary materials of C, titanium dioxide, and In_2O_3 are confirmed in the as-prepared nanocomposites.

2.2.3.3 Morphology and Structure Analysis

The surface and morphological characteristics of the prepared samples were observed by scanning electron microscopy (SEM) and illustrated in Fig. 2.19. There was no significant difference in the morphological characteristics of all carbon modified In_2O_3/titanium dioxide heterostructures. The aggregation of spherical nanoparticles can be observed from Fig. 2.19a. Compared with P25, the SEM images of C-P25 indicate that spherical nanoparticles are slightly dispersed. The same phenomena were observed in In/Ti-2 and C-In/Ti-2 samples. In addition, when the molar ratio of In/Ti exceeds 0.02:1, spherical nanoparticles become more and more aggregated with the increasing of In doping amount. The SEM image of In_2O_3 shows a cubic cube with condensed state. In order to further study the morphology and structure information of carbon modified In_2O_3/titanium dioxide heterostructures, the samples C-In/Ti-2 and

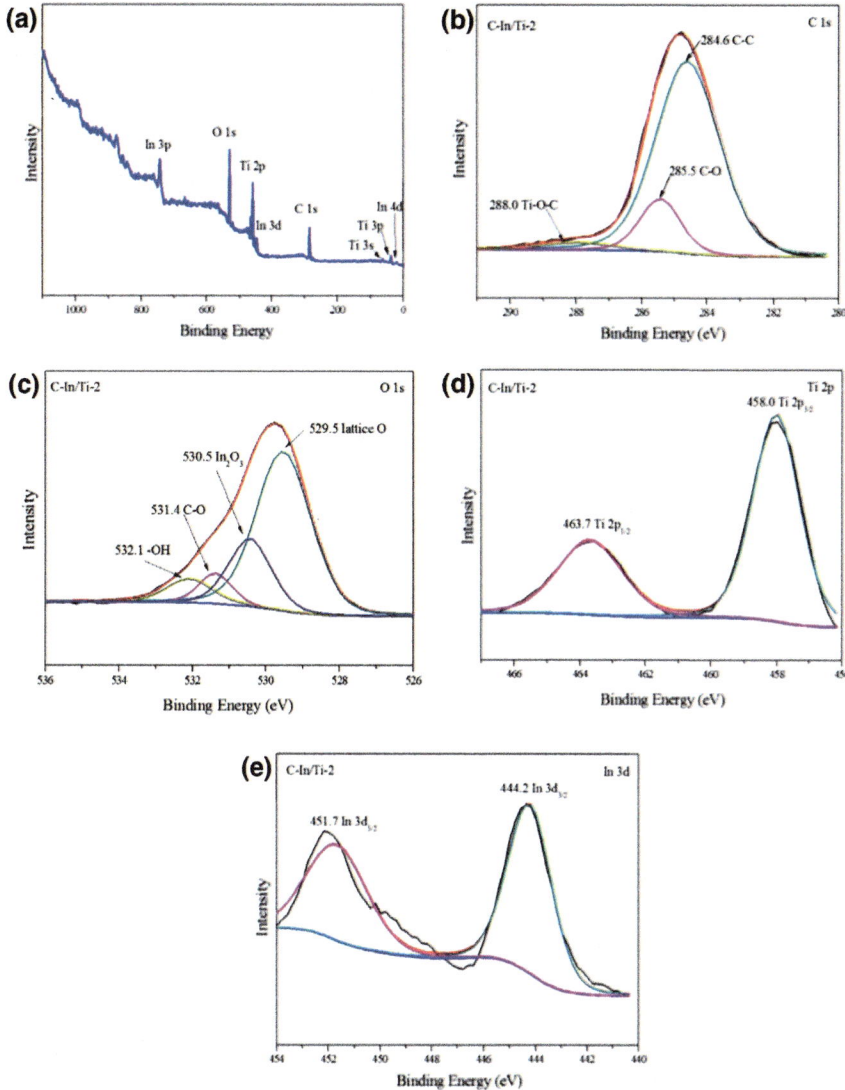

Fig. 2.18 XPS survey spectra of **a** C-In/Ti-2, high resolution XPS spectra of C-In/Ti-2 for **b** C 1s, **c** O 1s, **d** Ti 2p, **e** In 3d. Reprinted from Ref. [29], Copyright 2017, with permission from Elsevier

C-In/Ti-9 are analyzed by transmission electron microscopy (TEM). The results are shown in Fig. 2.19a, b. As can be seen from the TEM image, the samples of C-In/Ti-2 and C-In/Ti-9 are composed of approximately spherical nanoparticles between 15 and 35 nm, and the C-Si/Ti-2 samples are more dispersed than C-In/Ti-9 and are in agreement with the SEM images. At the same time, In_2O_3 nanoparticles are not observed in the TEM images of C-In/Ti-2 samples, indicating that In_2O_3 nanoparti-

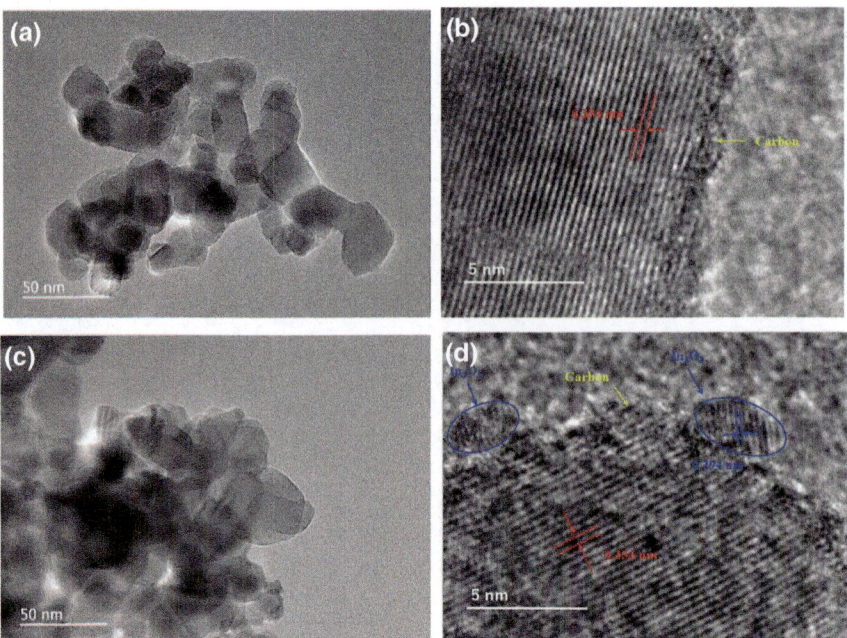

Fig. 2.19 TEM images of **a** C-In/Ti-2, **c** C-In/Ti-9, HRTEM images of **b** C-In/Ti-2, **d** C-In/Ti-9, **e** TEM and corresponding elemental mapping of C-In/Ti-2, **f** TEM and corresponding elemental mapping of C-In/Ti-9, EDS images of **g** C-In/Ti-2, **h** C-In/Ti-9. Reprinted from Ref. [29], Copyright 2017, with permission from Elsevier

cles are small and dispersed on the main surface, but in the C-In/Ti-9 samples, the particles are dispersed on the surface of titanium dioxide nanoparticles. However, the modified carbon was not found, so we identified the atomic morphology and structural information by high resolution transmission electron microscopy (HRTEM). HRTEM images of C-In/Ti-2 and C-In/Ti-9 samples are shown in Fig. 2.19c, d. As you can see from Fig. 2.19c, In_2O_3 nanoparticles are not found, which is consistent with the XRD pattern, and the modified carbon is wrapped on the surface of the titanium dioxide body. Meanwhile, the clear lattice stripe of 0.354 nm distance is allocated to the (101) surface of anatase titanium dioxide [31]. Moreover, as shown in Fig. 2.19d, modified carbon covering the surface of the main body is also found. The clear lattice fringes of 0.294 and 0.354 nm correspond to the (222) planes of cubic In_2O_3 and (101) anatase titanium dioxide respectively [32].

The spatial distribution information of Ti and In is analyzed by elemental map. As shown in Fig. 2.19e, f, the mapping of Ti and In elements is well defined and uniformly distributed in the whole nanocomposite, which confirms the successful preparation of In_2O_3/titanium dioxide nanostructures. At the same time, it confirms that In_2O_3 nanoparticles are very small and evenly dispersed on the main C-In/Ti-2 surface. The energy dispersive X ray spectroscopy (EDS) results of C-In/Ti-2 and

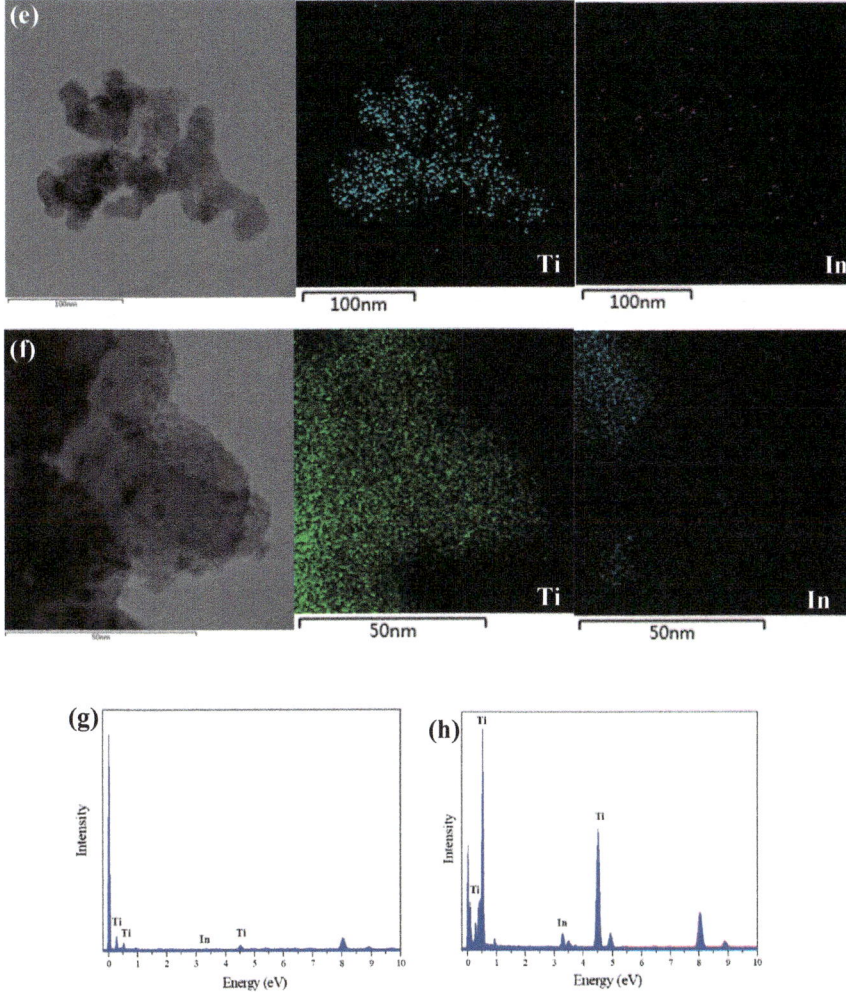

Fig. 2.19 (continued)

C-In/Ti-9 samples are shown in 2.19g, h. The percentages of Ti and In atoms of C-In/Ti-2 are 98.1% and 1.9% respectively, and the percentages of C-In/Ti-9 are 92.9% and 7.1% respectively, which are basically the same as the theoretical atomic ratio of In and Ti.

2.2.3.4 Surface Area Analysis

It is well known that the specific surface area (SBET) has a great influence on the photocatalytic reaction. N_2 adsorption/desorption measurements are used to deter-

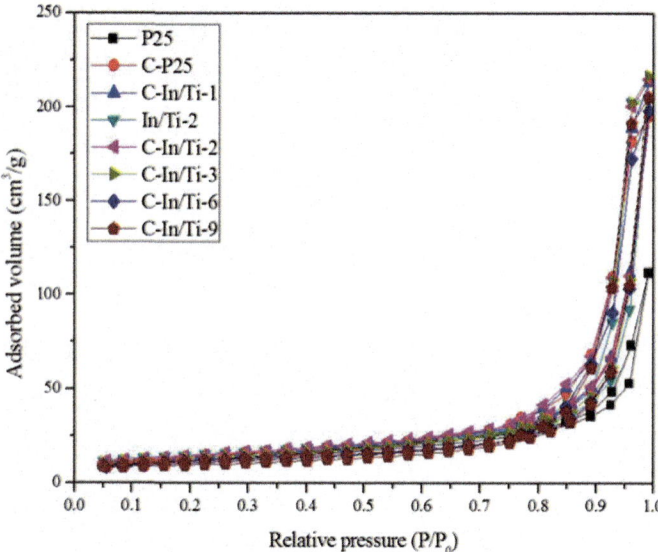

Fig. 2.20 N_2 adsorption-desorption isotherms of as-prepared samples. Reprinted from Ref. [29], Copyright 2017, with permission from Elsevier

mine the SBET of the prepared photocatalyst. The results are shown in Fig. 2.20. It is observed that all the isotherms are attributed to the typical IV type, and the H3 hysteresis loop [33] is in the high relative pressure range of 0.80–0.99. This means that all samples are mesoporous, which is due to the aggregation of titanium dioxide nanospheres, which is consistent with the pattern of SEM and TEM. Table 2.5 summarizes the SBET, pore volume and pore size. This shows that the use of carbon doped titanium dioxide can increase SBET, for P25, SBET increases from 41.86 m^2/g to 42.65 m^2/g, and In/Ti-2 increases from 43.91 m^2/g to 50.30 m^2/g. This may be due to the size reduction of nanoparticles and the inhibition of agglomeration after doping carbon. This means that doping carbon is a feasible way to improve the SBET of the main body. On the other hand, the SBET of carbon dioxide and indium oxide modified titanium dioxide is much larger than that of pure titanium dioxide. With the increase of In_2O_3 doping amount, C-In/Ti-2 samples have the largest SBET. It is generally believed that larger SBET usually produces better photocatalytic efficiency because photocatalytic reaction is a surface based phenomenon [34]. When the amount of indium oxide is lower than 1 mol%, the SBET of the sample increases with the increase of In_2O_3 doping amount. However, when the doping level of indium oxide exceeds 2 mol%, the opposite phenomenon occurs. This may be due to excess indium oxide hindering the pores and active sites of the main body, resulting in the decrease of the SBET of the three-component photocatalyst.

Table 2.5 The physical, chemical and photocatalytic efficiency of the as-prepared carbon modified In$_2$O$_3$/titanium dioxide heterostructure

Samples	BET surface area (m^2/g)	Pore volume (cm^3/g)	Pore diameter (nm)	Absorbing boundary (nm)	Bandgap (eV)	Hg^0 removal efficiency (visible light) η (%)
P25	41.86	0.173	16.56	386	3.21	17.5
C-P25	42.65	0.301	28.26	405	3.06	37.2
C-In/Ti-1	43.54	0.330	30.33	406	3.05	52.8
In/Ti-2	44.05	0.335	28.16	405	3.06	39.6
C-In/Ti-2	50.30	0.333	26.45	407	3.05	61.1
C-In/Ti-3	45.04	0.335	29.78	407	3.05	46.4
C-In/Ti-6	36.64	0.307	33.46	408	3.04	34.3
C-In/Ti-9	32.43	0.318	39.24	409	3.03	12.2
In$_2$O$_3$	–	–	–	470	2.64	8.1

Reprinted from Ref. [29], Copyright 2017, with permission from Elsevier

2.2.3.5 FT-IR Spectroscopy

Fourier transform infrared spectroscopy (FT-IR) is used to evaluate the bond characteristics of the official energy group in pure P25, In$_2$O$_3$, In/Ti-2 and C-In/Ti-X (X = 2, 9) (Fig. 2.21). In the FTIR spectra of In$_2$O$_3$ samples, three characteristic peaks of In$_2$O$_3$ appear. The peak at 433 and 563 cm^{-1} is In-O stretching vibration, and the peak at 601 cm^{-1} is [35] under In-O bending vibration. The peak at 1631 cm^{-1} and the broadband region near 3421 cm^{-1} correspond to the stretching vibration [36] of H-O-H and O-H in physisorption water. It is widely believed that one of the determinants of photocatalytic activity is the presence of the titanium dioxide surface hydroxyl group, which can capture a large number of OH radicals (H$^+$) and thus improve the photocatalytic activity. The broadband between 662 and 510 cm^{-1} is related to Ti-O-Ti stretching vibration.

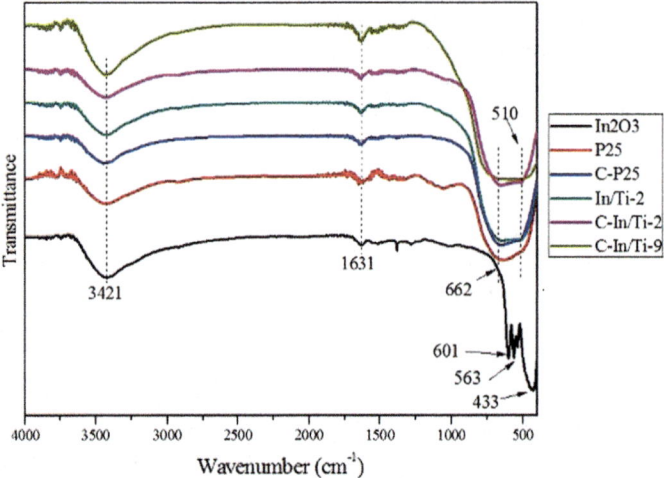

Fig. 2.21 The FTIR spectra of as-prepared samples. Reprinted from Ref. [29], Copyright 2017, with permission from Elsevier

References

1. J.M. Valero, S. Obregón, G. Colón, Active site considerations on the photocatalytic H_2 evolution performance of Cu-doped titanium dioxide obtained by different doping methods. Acs Catal. **4**(10), 3320–3329 (2014)
2. Xiao Zhou, Wu Jiang, Qifen Li, Yongfeng Qi, Zheng Ji, Ping He, Xuemei Qi, Pengfei Sheng, Qingwei Li, Jianxing Ren, Improved electron-hole separation and migration in V_2O_5/rutile-anatase photocatalyst system with homo-hetero junctions and its enhanced photocatalytic performance. Chem. Eng. J. **330**, 294–308 (2017)
3. J.S. Dalton, P.A. Janes, N.G. Jones, J.A. Nicholson, K.R. Hallam, G.C. Allen, Photocatalytic oxidation of NO_x gases using titanium dioxide: a surface spectroscopic approach. Environ. Poll. **120**, 415–422 (2002)
4. D. Sethi, N. Jada, A. Tiwari, S. Ramasamy, T. Dash, S. Pandey. Photocatalytic destruction of Escherichia coli in water by V_2O_5/titanium dioxide. J. Photochem. Photobiol B Biol. **144**, 68–74 (2015)
5. S. Hu, F. Zhou, L. Wang, J. Zhang, Preparation of Cu_2O/CeO_2 heterojunction photocatalyst for the degradation of acid orange 7 under visible light irradiation. Catal. Commun. **12**, 794–797 (2011)
6. C.C. Wang, J.R. Li, X.L. Lv, Y.Q. Zhang, G. Guo, Photocatalytic organic pollutants degradation in metal-organic frameworks. Energy Environ. Sci. **7**, 2831–2867 (2014)
7. D. Jiang, T. Wang, Q. Xua, D. Li, S. Meng, M. Chen, Perovskite oxide ultrathin nanosheets/g-C_3N_4 2D-2D heterojunction photocatalysts with significantly enhanced photocatalytic activity towards the photodegradation of tetracycline. Appl. Catal. B Environ. **201**, 617–628 (2017)
8. Y. Jin, D.L. Jiang, D. Li, M. Chen, Comparison of patterns and prognosis among distant metastatic breast cancer patients by age groups: a SEER population-based analysis. Catal. Sci. Technol. **7**, 2308–2317 (2017)
9. S.S.R. Putluru, L. Schill, A. Godiksen, R. Poreddy, S. Mossin, A.D. Jensen, R. Fehrmann, Promoted V_2O_5/titanium dioxide catalysts for selective catalytic reduction of NO with NH_3 at low temperatures. Appl. Catal. B Environ. **183**, 282–290 (2016)

10. S.S.R. Putluru, L. Schill, D. Gardini, S. Mossin, J.B. Wagner, A.D. Jensen, R. Fehrmann, Superior DeNOx activity of V_2O_5-WO_3/titanium dioxide catalysts prepared by deposition-precipitation method. J. Mater. Sci. **49**, 2705–2713 (2014)
11. C.Y. He, X. Li, Y.H. Li, J.F. Li, G.C. Xi, Large-scale synthesis of Au-WO_3 porous hollow spheres and their photocatalytic properties. Catal. Sci. Technol. **7**, 3702–3706 (2017)
12. B. Liu, Y. Xue, J. Zhang, B. Han, J. Zhang, X. Suo, L. Mu, H. Shi, Visible-light-driven titanium dioxide/Ag_3PO_4 heterostructures with enhanced antifungal activity against agricultural pathogenic fungi Fusarium graminearum and mechanism insight. Environ. Sci. Nano **4**, 255–264 (2017)
13. Q. Huang, S. Tian, D. Zeng, X. Wang, W. Song, Y. Li, W. Xiao, C. Xie, Enhanced photocatalytic activity of chemically bonded titanium dioxide/graphene composites based on the effective interfacial charge transfer through the C-Ti bond. Acs Catal. **3**, 1477–1485 (2013)
14. X. Yang, F.Y. Ma, K.X. Li, Y.N. Guo, J.L. Hu, W. Li, M.X. Huo, Y.H. Guo, Mixed phase titania nanocomposite codoped with metallic silver and vanadium oxide: new efficient photocatalyst for dye degradation. J. Hazard. Mater. **175**, 429–438 (2010)
15. R. Delaigle, P. Eloy, E.M. Gaigneaux, Necessary conditions for a synergy between Ag and V_2O_5 in the total oxidation of chlorobenzene. Catal. Today **175**, 177–182 (2011)
16. J. Wu, C. Li, X. Zhao, Q. Wu, X. Qi, X. Chen, T. Hu, Y. Cao, Photocatalytic oxidation of gas-phase Hg^0 by CuO/titanium dioxide. Appl. Catal. B: Environ. **176–177**, 559–569 (2015)
17. J. Zhong, F. Chen, J. Zhang, Carbon-deposited titanium dioxide: synthesis, characterization, and visible photocatalytic performance. J. Phys. Chem. C **114**, 933–939 (2010)
18. B. Li, Z. Zhao, F. Gao, X. Wang, J. Qiu, Mesoporous microspheres composed of carbon-coated titanium dioxide nanocrystals with exposed 001 facets for improved visible light photocatalytic activity. Appl. Catal. B: Environ. **147**, 958–964 (2014)
19. E.M. Neville, M.J. Mattle, D. Loughrey, B. Rajesh, M. Rahman, J.M.D. MacElroy, J.A. Sullivan, K.R. Thampi, Carbon-doped titanium dioxide and carbon, tungsten-codoped titanium dioxide through Sol-Gel processes in the presence of melamine borate: reflections through photocatalysis. J. Phys. Chem. C **116**, 16511–16521 (2012)
20. Y. Huang, W. Ho, S. Lee, L. Zhang, G. Li, J.C. Yu, Effect of carbon doping on the mesoporous structure of nanocrystalline titanium dioxide and its solar-light-driven photocatalytic degradation of NO_x. Langmuir **24**, 3510–3516 (2008)
21. S. Lee, Y. Lee, D.H. Kim, J.H. Moon, Carbon-deposited titanium dioxide 3D inverse opal photocatalysts: visible-light photocatalytic activity and enhanced activity in a viscous solution. ACS Appl. Mater. Interfaces. **5**, 12526–12532 (2013)
22. X. Yang, C. Cao, L. Erickson, K. Hohn, R. Maghirang, K. Klabunde, Photo-catalytic degradation of Rhodamine B on C-, S-, N-, and Fe-doped titanium dioxide under visible-light irradiation. Appl. Catal. B: Environ. **91**, 657–662 (2009)
23. Z. Ying, Z. Zhao, J. Chen, C. Li, J. Chang, W. Sheng, C. Hu, S. Cao, C-doped hollow titanium dioxide spheres: in situ synthesis, controlled shell thickness, and superior visible-light photocatalytic activity. Appl. Catal. B Environ. **165**, 715–722 (2015)
24. L.S. Yoong, F.K. Chong, B.K. Dutta, Development of copper-doped titanium dioxide photocatalyst for hydrogen production under visible light. Energy **34**, 1652–1661 (2009)
25. G. An, W. Ma, Z. Sun, Z. Liu, B. Han, S. Miao, Z. Miao, K. Ding, Preparation of titania/carbon nanotube composites using supercritical ethanol and their photocatalytic activity for phenol degradation under visible light irradiation. Carbon **45**, 1795–1801 (2007)
26. F. Dong, S. Guo, H. Wang, X. Li, Z. Wu, Enhancement of the visible light photocatalytic activity of C-doped titanium dioxide nanomaterials prepared by a green synthetic approach. J. Phys. Chem. C **115**, 13285–13292 (2011)
27. N.P. Zschoerper, V. Katzenmaier, U. Vohrer, M. Haupt, C. Oehr, T. Hirth, Analytical investigation of the composition of plasma-induced functional groups on carbon nanotube sheets. Carbon **47**, 2174–2185 (2009)
28. C. Chen, J. Moir, N. Soheilnia, B. Mahler, L. Hoch, K. Liao, V. Hoepfner, P. O'Brien, C. Qian, L. He, Morphology-controlled In_2O_3 nanostructures enhance the performance of photoelectrochemical water oxidation. Nanoscale **7**, 3683–3693 (2015)

29. Xiao Zhou, Wu Jiang, Qifen Li, Tao Zeng, Zheng Ji, Ping He, Weiguo Pan, Xuemei Qi, Chengyao Wang, Pankun Liang, Carbon decorated In_2O_3/titanium dioxide heterostructures with enhanced visible-light-driven photocatalytic activity. J. Catal. **355**, 26–39 (2017)
30. Z. Jiang, J. Zhu, D. Liu, W. Wei, J. Xie, M. Chen, In situ synthesis of bimetallic Ag/Pt loaded single-crystalline anatase titanium dioxide hollow nano-hemispheres and their improved photocatalytic properties. CrystEngComm **16**, 2384–2394 (2014)
31. M. Zeng, Y. Li, M. Mao, J. Bai, L. Ren, X. Zhao, Synergetic effect between photocatalysis on titanium dioxide and thermocatalysis on CeO_2 for gas-phase oxidation of benzene on titanium dioxide/CeO_2 nanocomposites. Acs Catal. **5**, 3278–3286 (2015)
32. Y. Meng, G. Liu, A. Liu, Z. Guo, W. Sun, F. Shan, Photochemical activation of electrospun In_2O_3 nanofibers for high-performance electronic devices. ACS Appl. Mater. Interfaces. **9**, 10805–10812 (2017)
33. M. Humayun, Y. Qu, F. Raziq, R. Yan, Z. Li, X. Zhang, L. Jing, Exceptional visible-light activities of titanium dioxide-coupled N-doped porous perovskite $LaFeO_3$ for 2,4-dichlorophenol decomposition and CO_2 conversion. Environ. Sci. Technol. **22**, 13600–13610 (2016)
34. M. Tahir, B. Tahir, N.A.S. Amin, H. Alias, Selective photocatalytic reduction of CO_2 by H_2O/H_2 to CH_4 and CH_3OH over Cu-promoted In_2O_3/titanium dioxide nanocatalyst. Appl. Surf. Sci. **389**, 46–55 (2016)
35. M. Epifani, R. Díaz, J. Arbiol, E. Comini, N. Sergent, T. Pagnier, P. Siciliano, G. Faglia, J.R. Morante, Nanocrystalline metal oxides from the injection of metal oxide sols in coordinating solutions: synthesis, characterization, thermal stabilization, device processing, and gas-sensing properties. Adv. Funct. Mater. **16**, 1488–1498 (2006)
36. R. Li, Z.J. Wang, L. Wang, B.C. Ma, S. Ghasimi, H. Lu, K. Landfester, K.A.I. Zhang, Photocatalytic selective bromination of electron-rich aromatic compounds using microporous organic polymers with visible light. Acs Catal. **6**, 1113–1121 (2016)
37. Z. Jin, W. Duan, W.B. Duan, B. Liu, X.D. Chen, F.H. Yang, J.P. Guo, Indium doped and carbon modified P25 nanocomposites with high visible-light sensitivity for the photocatalytic degradation of organic dyes. Appl. Catal. A **517**, 129–140 (2016)

Chapter 3
Preparation and Characterization Other Photocatalysts

Abstract A series of zinc-based photocatalyst and bismuth-based photocatalyst were prepared. Several preparation methods and their characteristics of zinc-based photocatalyst are given. $BiOIO_3$ nanosheets are studied systematically, which possess superior photocatalytic properties. The fabricated CSs-BiOI/$BiOIO_3$ composites with heterostructures show the enhanced photocatalytic performance than $BiOIO_3$. The physical and chemical properties of the as-prepared materials are characterized by X-ray diffraction, X-ray photoelectron spectroscopy, scanning electron microscope, transmission electron microscopy, high resolution TEM, Brunauer-Emmett-Teller, Fourier transform infrared, UV-vis diffuse reflectance spectra, and Photoluminescence.

Keywords Zinc-based · Bismuth-based · Nanosheets · Heterojunction

3.1 Zinc-Based Photocatalysts

The photocatalytic activity of ZnO semiconductor material is greatly influenced by morphology [1], particle size [2], and surface property [3]. Various synthetic methods of preparation of ZnO materials have been tested to try to improve the photocatalytic activity of ZnO semiconductors, such as solution method, hydrothermal/solvothermal growth, sol-gel method, chemical vapor deposition, and pyrolysis method [4, 5].

Xu et al. [6] prepared uniform flower-like ZnO superstructures on a large scale via a simple and inexpensive solution approach at room temperature. These ZnO microflowers having a mean diameter of about 4 μm, were assembled by numerous porous ZnO nanosheets with a thickness of 40 nm. The concentration of trisodium citrate and sodium hydroxide has great influence on the morphology of the final products respectively. At room temperature, the photoluminescence (PL) spectrum showed that the ZnO microflowers exhibited a very weak UV emission peak located at 380 nm and a relatively strong green emission peak centered at 565 nm.

Pál et al. [7] prepared ZnO and indium-doped ZnO structures with different morphologies by hydrothermal method at 150 °C. The structure formation was controlled by the zinc ion/hydroxide ion molar ratio and presence of l-histidine in the reaction mixture. TEM, FESEM measurements show that prism-like and flower-like crystals are formed during the synthesis, whose shape deformed due to the indium doping. XPS and EDX measurements confirm the incorporation of the dopant ion into the crystalline lattice of ZnO and zeta-potential investigations demonstrate the presence of indium ions on the surface of the particles. The indium ions-doping has a great influence on the optical properties (UV-diffuse reflection, fluorescence) of the particles. The bandgap energy of the samples decreases and the decrease becomes more significant with the increasing of indium concentration. The particles show visible emission in the region of 506–565 nm. The visible emission peak of the pure ZnO samples shifts toward the shorter wavelength due to the indium doping. Photoelectric properties of the doped samples were investigated using interdigitated microelectrodes. The results show that the photocurrent intensity decreases with the dopant concentration because of the presence of oxygen vacancies, which hinder the direct recombination of the photoexcited charge carriers.

Zhao et al. [8] prepared self-assembled 3D flower-like ZnO microstructures composed of nanosheets on a large scale through a sol–gel-assisted hydrothermal method using $Zn(NO_3)_2 \cdot 6H_2O$, citric acid, and NaOH as raw materials. The self-assembled 3D flower-like ZnO composed of nanosheets could be obtained over a relatively broad temperature range (90–150 °C) after 17 h of hydrothermal treatment. The sample prepared at 120 °C for 17 h exhibited superior photocatalytic activity to other ZnO samples and commercial ZnO, and it almost completely degraded a KGL solution within 40 min. The photocatalytic activity could attribute to the structure, surface defects, and surface areas of the samples.

Yang et al. [9] prepared highly efficient photocatalytic ZnO nanoneedle arrays with a large surface/volume ratio on inexpensive, large-area substrates using metal-organic chemical vapor deposition. The photocatalytic activity of ZnO nanoneedle arrays is much enhanced due to their increased surface/volume ratio. It is believed that the 'bottom-up' approach may be expanded to create many other one-dimensional oxide semiconductor nanostructures.

Parhizkar et al. [10] prepared nanocrystalline, zinc oxide films by the pyrolysis of zinc arachidate LB multilayers in oxygen ambient. FT-IR and UV-vis spectroscopy showed that on heat treatment, the pyrolysis began at ~200 °C and was complete at ~350 °C. TEM studies show that the heat treatment results in the formation of a continuous and uniform ZnO film consisting of nanocrystallites of size 5 ± 2 nm. Electron diffraction studies show that the film consists of an unusual phase reported as "leafy and thin platy crystal", obtained under high-pressure conditions.

The band gap of the film, estimated from a plot of α versus $h\upsilon$, was found to be ~3.3 eV. It is interesting to note that the presence of the unusual phase and nanocrystalline nature do not have a significant effect on the optical gap of the ZnO film.

3.2 Bismuth-Based Photocatalysts

3.2.1 Introduction of Bi-Based Photocatalysts

In order to improve utilization of solar energy, the researchers explored a large number of new photocatalysts, in which bismuth oxyhalides (BiOX, X=F, Cl, Br, I) is such a new type of semiconductor. BiOX crystal structure is a tetragonal lattice type, which is composed of double halogen atomic layer and $(Bi_2O_2)^{2+}$ layer to alternately form layer structure. Thus BiOX compounds tend to grow into nanosheets/nanoplates with 2D structures. Compared to two-dimensional (2D) nanosheets, the crystal morphology of three-dimensional (3D) hierarchical structures can provide novel or enhanced properties, including stronger light absorption, more reactive points, faster separation of interfacial charge and shortened diffusion pathways, which could improve photocatalytic performance. In recent years, scientists have used different synthetic methods to synthesize 3D BiOX microstructures with different morphologies, such as microflowers, microspheres and so on. Yan Mi et al. synthesized novel 3D graphene-like ultrathin BiOCl hierarchical architectures using a surfactant assisted hydrothermal process and the resultants showed a superior photocatalytic performance of 99% photodegradation and corresponding 74% mineralization for rhodamine B within 10 min, and an outstanding photocurrent response under solar light illumination. Liyong Ding et al. synthesized uniform BiOCl hierarchical microspheres assembled by nanosheets with tunable thickness via a simple solvothermal route and the specific surface area of BiOCl hierarchical microspheres increased, which was favorable to photocatalytic activities for the RhB degradation under visible light irradiation. The above researches were to prepare the photocatalysts by template methods such as introducing template agents and surfactants in the preparation process, which is more cumbersome and costly, limiting its application. In contrast, no template method is without the introduction of templating agents and other surfactants so that it is inexpensive and, environmentally friendly and has favorable synthesis conditions, which is suitable for large scale applications.

In addition to controlling the morphology of photocatalysts, combining BiOX with other semiconductors to construct heterojunctions is an effective way to develop excellent photocatalysts. Compared to the single semiconductor photocatalyst, the interfacial contact electric field in the heterogeneous structure photocatalyst could promote the effective separation of the photo-generated electron-hole pairs. In addition, the formation of heterojunction can adjust to the band structure, so that it has a more suitable band position, which could reduce the recombination rate of electron hole pairs. Jinfeng Zhang et al. developed a simple synthetic approach of nanosheet-assembled BiOCl/BiOBr microspheres by an ethylene glycol (EG)-assisted hydrothermal method, and the efficient separation of e^- and h^+ originated from the formation of BiOCl/BiOBr heterojunction interface was benefitul to obtain higher photocatalytic activity. Qin et al. facilely prepared flower-like $BiOCl_{1-x}Br_x$ hierarchical microspheres assembled by nanosheets through solvothermal treatment, the balance between suitable band gap and adequate potential of valence band in BiOCl$_{1-x}$Br$_x$ crystals dominated their photocatalytic activity.

All the BiOX photocatalysts have tetragonal structure, which is composed of [Bi$_2$O$_2$] slabs overlapping with halogen atoms. As is well known, exposed facet is one of factors affecting the photocatalytic performance of semiconductor. Ye et al. synthesized BiOI single-crystal nanosheets with dominant exposed {001} facets (up to 95%). Pan et al. reported that exposed (110) facets was greater than exposed (001) facets in the photo-degradation of bisphenol A. Sun et al. found the BiOI(001)/BiOCl(010) heterojunctions displayed enhanced visible-light photocatalytic activity. The above preparation adopted calcination method, hydrothermal method and solvent heat method. These synthesis methods are complicated and costly, so room temperature synthesis is a feasible and widely applicable method. Moreover, the microstructures and even the chemical and physical properties of semiconductor surfaces and interfaces can be controlled by various parameters in the synthesis process. Synergistic effects of different crystal surfaces of semiconductor has a positive effect on the photocatalytic activity.

Bismuth oxyhalides (BiOX; X=Cl, Br, I) are an attractive semiconductors in environmental remediation and solar energy conversion because of its unique V-VI-VII ternary compound, exhibiting a layered-structure induced fascinating physicochemical properties and suitable band-structure, as well as their high chemical stability, nontoxicity and corrosion resistance. As a tetragonal matlockite (PbFCl-type) structure characterized by {Bi$_2$O$_2$} slabs and interleaved by double slabs of halogen atoms, exhibited an excellent photogenerated electrons (e$^-$) holes (h$^+$) pairs separation process. The e$^-$ and h$^+$, generated by the visible light process, are located at the active sites of the nanocrystal surface, where they act as reduction sources or oxidation sources, leading to the produce reactive oxygen species (ROS). The ROS just like superoxide radical ($\cdot$O$_2^-$) and free hydroxyl groups ($\cdot$OH) are the dominated oxidative radicals to degrade the organics and oxide the inorganic matters in the homogeneous catalysis or heterogeneous catalysis. In these studies, the researchers focused on the morphology control, forming the heterostructure, in situ ions doping, introducing intrinsic oxygen vacancy, and exposed reactive facets to enhance the photocatalytic activity of the Bismuth oxyhalides. Among the Bismuth oxyhalides, the BiOI, which is a kind of Bismuth oxyhalide, has received more attention due to its narrow band gap (1.75 eV) and can be excited under visible light directly. At the same time, the photocatalytic activity deeply depends on the selectively exposed facets, because the photogenerated e$^-$ and h$^+$ migrate to different facets exhibiting discrepancy photocatalytic activities. Furthermore, the exposed different facets may induce the nanocrystal a different atomic arrangement and electronic properties, resulting in different physicochemical properties.

Yang et al. made a breakthrough in anatase titanium dioxide with exposed (001) facets by fluorine-terminated surfaces method, and has found the titanium dioxide with exposed (001) facets are more energetic than the thermodynamically stable (101) facets. Soon afterwards, the influences of nanocrystal semiconductor with exposed different facets were investigated and published. Massimiliano et al. reported that the titanium dioxide with exposed (001) can enhance the photocatalytic activity because of the accumulation of the trapped holes, however, the titanium dioxide with exposed (101) facets can increase the concentration of Ti^{3+} centers, resulting in worst photooxidative efficacy. Li et al. found that charge separation can be achieved by the photogenerated e^- and h^+ selectively located on the (010) and (110) facets of $BiVO_4$ respectively, which attain a higher activity in both photocatalytic and photoelectrocatalytic water oxidation reactions. These properties can be attributed to the different atomic arrangements on the different exposed facets, which exhibited different conduction and valence band energy level, resulting in the selectively photogenerated e^- and h^+ seperation process.

For bismuth oxyhalides, Jiang et al. investigated that the BiOCl with exposed (001) facets obtained a higher activity on photoexcitation pollutant degradation due to the synergistic effect between the surface atomic structure and suitable internal electric fields. Conversely, the BiOCl with exposed {010} facets processed larger surface area and open channel characteristics, being with superior activity on indirect dye photosensitization degradation. Pan et al. reported that the BiOI with exposed (110) and (001) facets can be synthesized by hydrothermal method at a different reaction time, and the results illustrated that BiOI-110 (BiOI-110 represented the BiOI with exposed 110 facets) can produce more $·O_2^-$ than BiOI-001 (BiOI-001 represented the BiOI with exposed 001 facets). Furthermore, the BiOI-110 can generate $·OH$, but the BiOI-001 cannot generate $·OH$. On the other hand, Ye et al. also fabricated the BiOI with exposed (100) and (001) facets by hydrothermal method at a different reaction time, but found the BiOI-001 had an excellent performance on photogenerated e^- and h^+ than BiOI-100, and were of a great photocatalytic CO_2 conversion efficiency. It seems that there are some contradictory descriptions on the BiOI with exposed different facets exhibiting the different efficacy on the pollution control or energy conversion due to the insufficiently systematical characterization.

3.2.2 Characterization of $BiOIO_3$ Nanosheets

3.2.2.1 Preparation of Samples

$BiOIO_3$ was prepared by a simple hydrothermal method. In a typical procedure, 0.485 g of $Bi(NO_3)_3 \cdot 5H_2O$ was dissolved into 80 mL H_2O, and stirred vigorously for 30 min. Then, 0.214 g of KIO_3 was added into the above aqueous and continuously stirred for 10 min. The obtained suspension was then hydrothermally treated at 150 °C for 6 h. Finally, the $BiOIO_3$ product was collected and dried at 60 °C for 12 h.

$BiOI/BiOIO_3$ heterostructured nanocomposites were synthesized by a chemical precipitation method at room temperature. A certain amount of $Bi(NO_3)_3 \cdot 5H_2O$ was dissolved in 80 mL H_2O containing 9 mL of acetic acid, and 0.48 g of $BiOIO_3$ was added into the above aqueous solution and stirred for 10 min. Then, 30 mL aqueous solution containing stoichiometric amount of KI was added dropwise into the solution and stirred for 2 h. After the stirring was completed, the resulted suspension was aged for 2 h. Finally, the resulted products were collected by filtration, washed with water and ethanol for several times and dried at 60 °C to obtain the final products. Depending on the molar ratio of BiOIO$_3$ to BiOI (1:0, 6:1, 3:1, 1:1, 0:1), different nanocomposites can be synthesized and named as BiOIO$_3$, B-6, B-3, B-2, B-1 and BiOI respectively.

3.2.2.2 XRD Analysis

The XRD patterns of the as-prepared BiOI/BiOIO$_3$ catalysts were depicted in Fig. 3.1. The XRD diffraction peaks of BiOIO$_3$ were in good agreement with the orthorhombic BiOIO$_3$ (ICSD # 262019). The obtained BiOI sample was well crystallized and can be indexed to tetragonal structure for BiOI (JCPDS file no. 73-2062). The (010) and (040) peaks of BiOI/BiOIO$_3$ heterostructures offset to a large angle to a certain extent compared to that of the pure BiOIO$_3$, which is because the samples shifted towards the long wavelength. The enlarged spectrum of (010) peaks were shown in Fig. 3.1b, and the intensity of (010) peak first gradually increased from B-6 to B-3 and then decreased from B-3 to B-1 with the increase of the content of BiOI, i.e., the intensity of (010) peak was the strongest when the mole ratio of BiOI/BiOIO$_3$ was 3:1. We can see from the enlarged XRD spectrum (Fig. 3.1c), the intensity of (040) peak of BiOIO$_3$ was also the strongest when the mole ratio of BiOI/BiOIO$_3$ compounds was 3:1. The Fig. 3.1d revealed the ratio of (040)/(002) of BiOI/BiOIO$_3$ compounds, which showed the same trend that B-3 possessed the largest intensity ratio value of (040)/(002). It verified that B-3 possesses the highest exposure {010} facet, while excessive BiOI products may cover the {010} facet of BiOIO$_3$, conversely reducing the exposure ratio [11, 12]. Therefore the B-3 sample may have the highest photocatalytic activity.

3.2.2.3 Optical Properties

The optical properties of the BiOI/BiOIO$_3$ compounds were confirmed by UV-vis diffuse reflectance spectroscopy. Figure 3.2a revealed that the as-prepared BiOI/BiOIO$_3$ compounds owned an excellent visible light absorption performance, while the pure BiOIO$_3$ and BiOI can be excited before 380 nm and 700 nm respectively. For a semiconductor, the band gap energy is described by the following equation [13]:

$$\alpha h\upsilon = A(h\upsilon - E_g)^{n/2} \quad (3.1)$$

where α is the absorption coefficient, h is the photon energy, A is a constant and E_g is the band gap. Both of BiOIO$_3$ and BiOI are indirect transition semiconductor, so for them, n = 4. From the $(\alpha h\upsilon)^{1/2}$ versus $h\upsilon$ plot (Fig. 3.2b), the band gap energies of BiOIO$_3$ and BiOI are 3.13 and 1.75 eV respectively. In the same way, the band gap energies of B-6, B-3, B-2 and B-1 are 3.04, 2.72, 2.61 and 1.86 eV respectively. The valence band-edge potential of a semiconductor can be calculated by the empirical equation $E_{VB} = \chi - E^e + 0.5E_g$ [14], where χ is the electronegativity of the semiconductor atoms, E^e is the energy of free electrons on the hydrogen scale

3.2 Bismuth-Based Photocatalysts

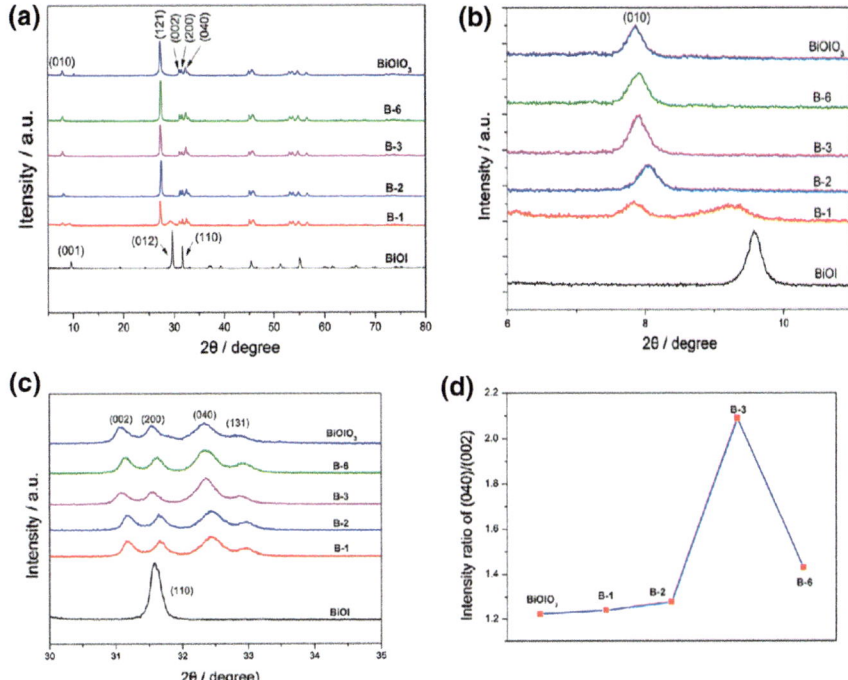

Fig. 3.1 XRD pattern of the BiOIO$_3$, BiOI and BiOI/BiOIO$_3$ heterostructures. Reprinted from Ref. [11], Copyright 2017, with permission from Elsevier

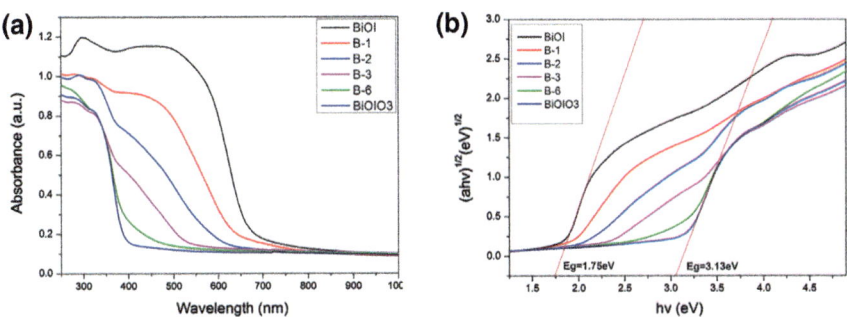

Fig. 3.2 a UV-vis DRS of the as-prepared samples and b band gap of BiOIO$_3$, BiOI and BiOI/BiOIO$_3$ heterostructures. Reprinted from Ref. [11], Copyright 2017, with permission from Elsevier

(about 4.5 eV), and E_g is the band gap energy of semiconductor. The CB bottom E_{CB} can be determined by $E_{CB} = E_{VB} - E_g$. The χ value of BiOIO$_3$ is about 7.04 eV, so the E_{VB} is calculated to be 4.11 eV, and E_{CB} is calculated to be 0.98 eV. For BiOI, the value is about 5.99 eV, hence the E_{VB} is calculated to be 2.37 eV, and E_{CB} is calculated to be 0.62 eV.

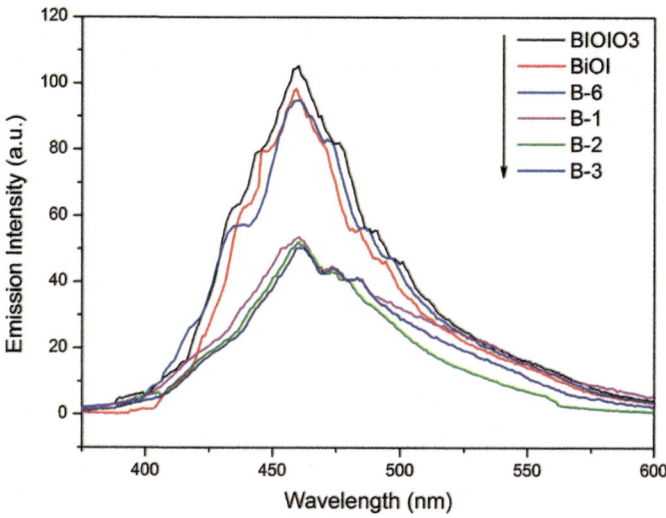

Fig. 3.3 Photoluminescence spectra of BiOIO$_3$, BiOI and BiOI/BiOIO$_3$ compounds. Reprinted from Ref. [11], Copyright 2017, with permission from Elsevier

3.2.2.4 Photoluminescence Spectra

PL is a powerful characterization method for detecting the separation efficiency of e^--h^+ [15, 16]. The lower intensity of the PL peak, the higher separation efficiency of e^--h^+ [15, 16]. With an excitation wavelength at 315 nm, the PL spectra of the as-prepared samples was presented in Fig. 3.3. It can be seen that all samples had emission peaks at about 460 nm, and the emission peak strengths of BiOI/BiOIO$_3$ compounds were lower than that of the pure BiOIO$_3$ and BiOI, which demonstrated that the BiOI/BiOIO$_3$ compounds were favorable for the separation of the e^--h^+. Among all the emission peaks, the emission peak of B-3 was the weakest, so it can be speculated that the photocatalytic activity of B-3 was the best.

3.2.2.5 BET Analysis

The specific surface area is usually closely related to the activity of photocatalyst, and generally the larger the specific surface area, the more active sites exposed, the more conducive to the photocatalytic reaction. The specific surface areas and Hg^0 removal efficiencies of the as-prepared samples were listed in Table 3.1. The specific surface areas of BiOI/BiOIO$_3$ compounds were almost the same sizes, but the Hg^0 removal efficiencies of BiOI/BiOIO$_3$ compounds differed greatly. The B-3 had the highest photocatalytic activity, whereas the specific surface area was not the largest, so the specific surface areas of BiOI/BiOIO$_3$ compounds were not the major factor that affects the photocatalytic properties. We assumed the exposed reactive facets of

3.2 Bismuth-Based Photocatalysts

Table 3.1 The BET surface areas and the Hg^0 removal efficiency of the samples

Products	BET surface area (m²/g)	Removal efficiency (LED) (%)
BiOI	2.1517	12.45
B-1	14.4314	61.35
B-2	15.9582	80.71
B-3	14.2358	98.53
B-6	13.4529	91.95
$BiOIO_3$	27.7027	59.31

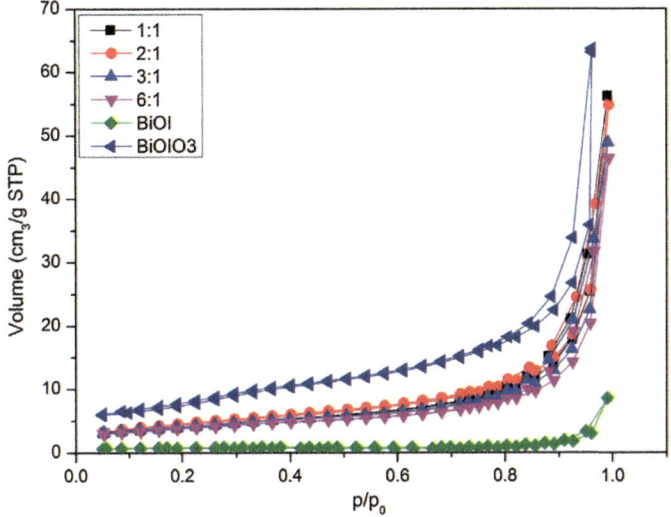

Fig. 3.4 Nitrogen adsorption-desorption isotherms. Reprinted from Ref. [11], Copyright 2017, with permission from Elsevier

the (010) facets of $BiOIO_3$ and the {001} facets of BiOI were the dominant factors that affect the photocatalytic properties. Figure 3.4 revealed that nitrogen adsorption-desorption isotherms were all IV type and the existed hysteresis loop extended from $P/P_0 = 0.6$ to $P/P_0 = 1$, demonstrating that the as-prepared $BiOI/BiOIO_3$ compounds were mesoporous and macroporous materials, and that the pore structure was mainly formed by the stacking of nanosheets, which was consistent with the SEM and TEM results.

3.2.2.6 BiOI/BiOIO₃ SEM and TEM Analysis and Its Formation Mechanism

The morphologies of $BiOIO_3$, BiOI and $BiOI/BiOIO_3$ compounds were explored by SEM and TEM. Figure 3.5a showed that BiOI approximated to rectangular nanosheet with smooth surface. Figure 3.5b revealed that the structure of $BiOIO_3$ was also smooth nanosheet. Figure 3.5c and d were respectively corresponding to the low and high magnification SEM spectrum of B-3, which indicated the $BiOIO_3$ and BiOI nanosheets stacked together in a regular and repeating pattern. Figure 3.5e and f verified the neat and smooth surfaces of the pristine $BiOIO_3$ and BiOI, and the diameters of $BiOIO_3$ and BiOI were 100 and 200 nm respectively. The lattice fringe can be clearly seen in HRTEM spectrum through Fig. 3.5g and h, and the lattice fringes with the interplane space of 0.287 nm corresponded to (002) plane of $BiOIO_3$. Combined with the XRD spectrum, it can be explored that the dominant exposed reactive facet of $BiOIO_3$ may be {010} facet. The lattice fringes of the interplane space of 0.280 nm corresponded to the (110) plane of BiOI. According to the (001) peak of XRD, the major exposed reactive facet of BiOI was {001} facet. The exposed {001} facets of BiOI can promote the photocatalytic activity, which was due to the effective separation of photogenerated e^--h^+ on {001} facets [12]. The BiOI were formed on the surface of $BiOIO_3$, which indicated that the BiOI/$BiOIO_3$ heterostructures were synthesized and the BiOI/$BiOIO_3$ layers contacted intimately.

The BiOI/$BiOIO_3$ formation mechanism was shown in Fig. 3.5i. The {010} facets of $BiOIO_3$ were first synthesized. The {010} facets of $BiOIO_3$ were formed by the oxygen atoms, and the $(Bi_2O_2)^{2+}$ layer connected the $(IO_3)^-$ layer to generate $BiOIO_3$. As the atoms arrangement of {001} facets of BiOI was similar to that of the {010} facets of $BiOIO_3$, the {001} facets of BiOI grew along the {010} facets of $BiOIO_3$, and the I slabs with the $[Bi_2O_2]$ sheets stacked together to form $[Bi_2O_2I_2]$ by nonbonding interaction [17]. Due to the sharing of oxygen atoms of {010} facets and {001} facets, which can increase contacting areas and intimate interface, the distinctive structure can effectively promote the transfer of photogenerated e^--h^+ between the BiOI/$BiOIO_3$ heterostructures [17].

3.2.2.7 XPS Analysis

The surface compositions of the samples were analyzed by X-ray photoelectron spectroscopy (XPS), shown as Fig. 3.6. It can be seen from Fig. 3.6a that the as-prepared samples contained the elements of I, O, Bi and C. The peak for C 1s(284.8 eV) was just attributed to the adventitious carbon. Figure 3.6b indicated that B-3 displayed two sets of I 3d peaks, and the two strong peaks at around 638.4 and 627.2 eV were assigned to I $3d_{5/2}$ and I $3d_{3/2}$ states of I^{5+}, whereas the other two strong peaks at about 633.2 and 621.4 eV were attributed to I $3d_{3/2}$ and I $3d_{5/2}$ states of I^-, and the I^{5+} and I^- ions were in $BiOIO_3$ and BiOI respectively. The Bi atoms exhibited the peaks at 164.8 and 158.8 eV (Fig. 3.6c), corresponding to Bi $4f_{2/5}$ and Bi $4f_{2/7}$, which indicated that the Bi atoms in B-3 were in the form of Bi^{3+}. Moreover, the O 1s region can be indexed to the peak at 532.8 eV (Fig. 3.6d), and the peak belonged to O^{2-} in Bi-O bands and I-O bands. Compared with the pristine $BiOIO_3$ and BiOI, the binding energy of I, Bi and O in BiOI/$BiOIO_3$ heterostructures (B-3) all shift towards high binding energy, which indicated that the interaction between $BiOIO_3$ and BiOI at the composite interface did occur, and the XPS spectrum can also confirm

3.2 Bismuth-Based Photocatalysts

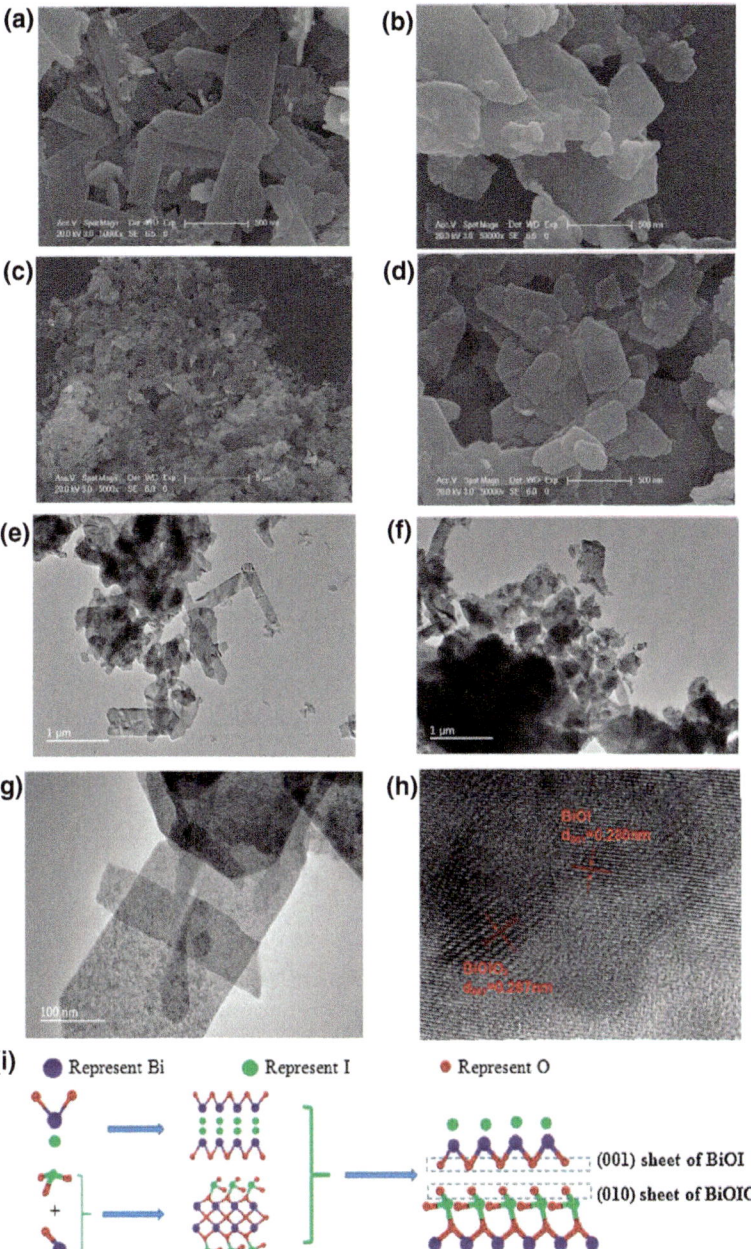

Fig. 3.5 The SEM images of BiOI (**a**); BiOIO$_3$ (**b**); B-3 (**c, d**); TEM images of BiOIO$_3$ and BiOI (**e, f**); HRTEM images of B-3 (**g, h**) and the formation mechanism (**i**). Reprinted from Ref. [11], Copyright 2017, with permission from Elsevier

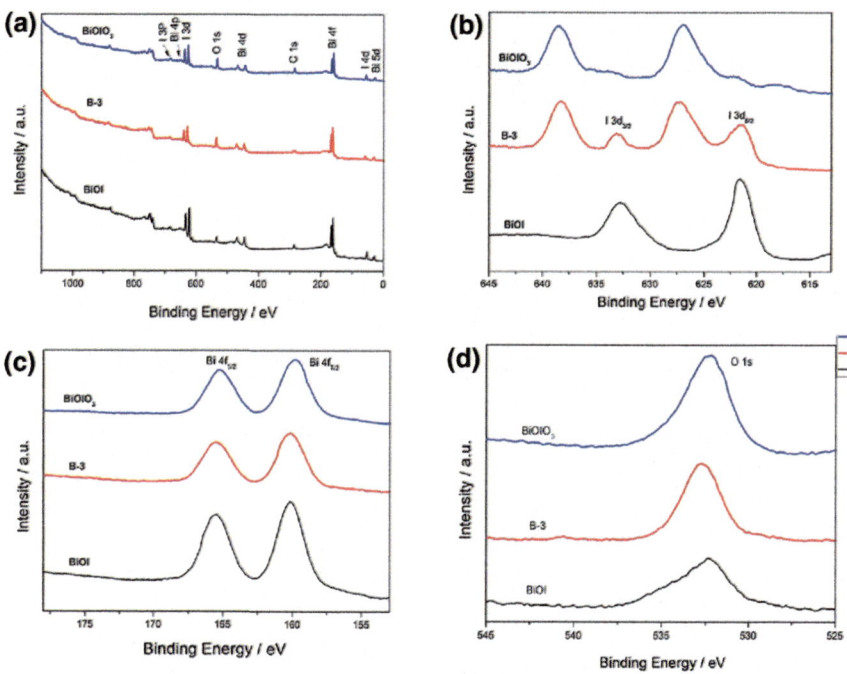

Fig. 3.6 XPS spectrum of BiOIO$_3$, BiOI and the B-3: survey XPS spectrum (**a**), I 3d (**b**), Bi 4f (**c**) and O 1s (**d**). Reprinted from Ref. [11], Copyright 2015, with permission from Elsevier

the coexistence of BiOIO$_3$ and BiOI in BiOI/BiOIO$_3$ composites and the effective composites were formed between the two kinds of the pristine phase materials.

3.2.3 Characterization of CSs-BiOI/BiOIO$_3$ Composites with Heterostructures

3.2.3.1 Synthesis of CSs-BiOI/BiOIO$_3$

The synthesis process of CSs-BiOI/BiOIO$_3$ was as follows: a certain quantity of CSs was added into 80 mL distilled water, and the suspension was sonicated for 30 min to completely disperse the CSs. Then, 0.485 g Bi(NO$_3$)$_3$ · 5H$_2$O was dissolved in the CSs suspension, and then it was constantly stirred for 20 min. 0.214 g KIO$_3$ was added into the above mixture, followed by stirring for 10 min. The resulting aqueous suspension was transferred into 100 mL Teflon-lined stainless autoclave, which was heated to 150 °C for 6 h. After naturally cooling down to room temperature, the CSs-BiOI/BiOIO$_3$ products were obtained by filtering and washed for three times with deionized water and ethanol, and then dried at 60 °C for 12 h. The CSs-BiOI/BiOIO$_3$

nanocomposites with different weight ratios of CSs to BiOIO$_3$ (0.5, 1, 3, 5, 10, 15 and 25 wt%) were denoted as CSs-BOI-0.5, CSs-BOI-1, CSs-BOI-3, CSs-BOI-5, CSs-BOI-10, CSs-BOI-15 and CSs-BOI-25 respectively. As a reference, we prepared pure BiOIO$_3$ without CSs and with other conditions remaining the same and denoted it as P-BiOIO$_3$.

3.2.3.2 Crystal Structure of CSs-BiOI/BiOIO$_3$ Heterojunctions

The XRD patterns of as-prepared CSs-BiOI/BiOIO$_3$ samples are shown in Fig. 3.7. The peak at 21.04° in the pure CSs sample can be corresponded to the (002) diffraction modes of the graphitic structure, which is the characteristics of disordered carbon materials. The diffraction peaks of BiOIO$_3$ and CSs-BiOI/BiOIO$_3$ are indexed to orthorhombic BiOIO$_3$ (ICSD # 262019, space group:Pca21; a = 5.6584(4), b = 11.0386(8), c = 5.7476(4) Å), and no typical peaks belonging to CSs are found presumably due to the low content of CSs in the nanocomposites and the peak intensities are too low comparing to that of BiOIO$_3$. As the weight ratio of CSs increases from 0 to 10, all diffraction peaks decrease gradually, suggesting that the CSs would affect the crystallinity of BiOIO$_3$ or the hybrid structures of amorphous carbon and crystalline BiOIO$_3$. In the enlarged image (Fig. 3.7b), according to tetragonal phase (JCPDS No. 10-0445), it can be seen that the diffraction peaks of BiOI appear while the CSs weight ratio is more than 5, and the peaks increase with the CSs increasing. After the pH detection, we have found that the pH of solution 1 and 2 in Fig. 3.8 are about 2.5 and 3 respectively. It shows that the hydrogen ion is consumed in the hydrothermal process with the CSs adding. As everyone knows, starch solution turns into blue when it encounters iodine. In Fig. 3.8, for the starch solution 1 + 3, no color change while the starch solution 2 + 4 turned into dark blue. The strong evidence can prove that after hydrothermal treatment with CSs there is iodine produced. Iodine is dissolved in ethanol and washed so that the final powders do not contain iodine. We can account that in the presence of high temperature, high pressure and acid condition, the BiOIO$_3$ precursor reacts with CSs and produces BiOI/BiOIO$_3$ solid and iodine.

Firstly, Bi(NO$_3$)$_3$ was hydrolyzed in deionized water, and reacted with KIO$_3$ to produce BiOIO$_3$ crystals. In the hydrothermal process, the BiOIO$_3$ grew on the integrated CSs, and the IO$_3^-$ is reduced by C into I$^-$ and I$_2$, producing CO$_2$ and H$_2$O. The formed BiOI will locate on the surface of BiOIO$_3$ and the CSs-BiOI/BiOIO$_3$ heterostructure was synthesized. With increasing CSs amount, this can promote. Ultimately, majority of BiOIO$_3$ would transform to BiOI, there is a small quantity of BiOIO$_3$ existed in CSs-BiOI/BiOIO$_3$ heterostructure, and this phenomenon can be seen from Fig. 3.9b. Figure 3.9 demonstrates the SEM images of CSs and CSs-BiOI/BiOIO$_3$ composites by the hydrothermal method. As shown in Fig. 3.9a, the average size of the pristine CSs is about 1 lm and they have uneven surfaces, and the surface roughness of CSs can be observed. Figure 3.9b displays the SEM images of the pure BiOIO$_3$ sample, and we can see that pure BiOIO$_3$ are composed of nanoplates and nanoparticles with different sizes. Layered structure on its micro-

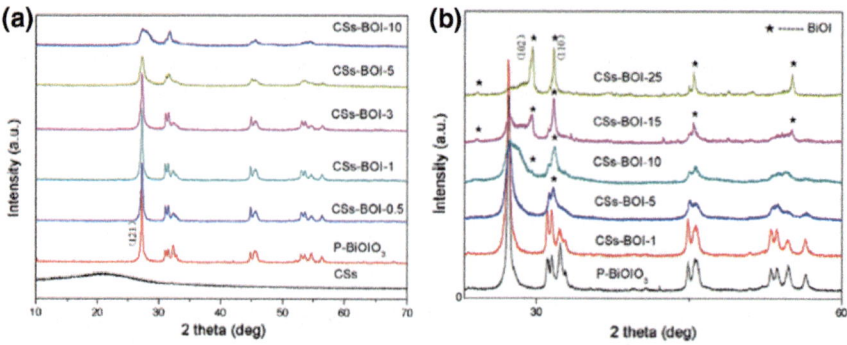

Fig. 3.7 XRD patterns of CSs, BiOIO₃ and CSs-BiOI/BiOIO₃ (**a**) and partial magnified details (**b**). Reprinted from Ref. [18], Copyright 2016, with permission from Elsevier

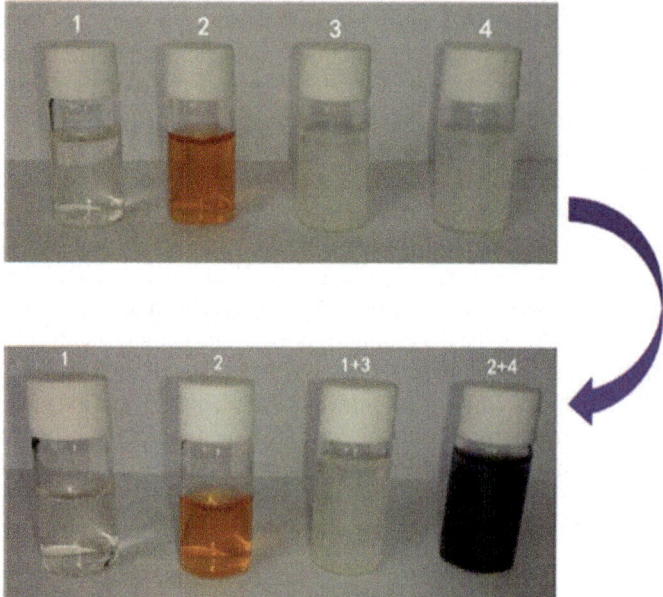

Fig. 3.8 Iodine detection, solution 1 is the solution after hydrothermal reaction inpreparation of P-BiOIO₃, solution 2 is the solution after hydrothermal reaction inpreparation of CSs-BOI-10, both solution 3 and 4 are the starch solution. Reprinted from Ref. [18], Copyright 2016, with permission from Elsevier

scopic morphology determines their macroscopic crystal flake. After hydrothermal treatment with CSs, the nanoplates became small and nanoparticles were increased, and this can result in the enormous surface area, which is consistent with the BET result. No carbon spheres can be found when the amount of CSs is very low(weight ratios = 0.5%), which can be attributed that the growth of $BiOIO_3$ disrupts or covers the structure of CSs. Besides, some intact CSs and more nanoparticles are detected with increasing CSs content. This can be due to $BiOIO_3$ growing on the integrated CSs and the ruinate CSs attaching on the surface of $BiOIO_3$ so that they in turn prevent the growth of the nanosheets. Furthermore, the formed BiOI locates on the surface of $BiOIO_3$, in agreement with the XRD analysis, which also affects the morphology of $BiOIO_3$.

3.2.3.3 Morphology and Structure

Figure 3.9a and b show the TEM and HRTEM images of CSs-BOI-5composite. It can be found that the lattice fringes are composed of two parts. One is observed with fringe spaces of 0.282 nm, which is corresponding to the (102) plane of tetragonal BiOI. The other lattice fringe has a spacing of 0.324 nm, being in well agreement with the spacing of the (121) plane of $BiOIO_3$. The aforesaid lattice fringes conform to the XRD peaks of the samples. In such a manner, the BiOI and $BiOIO_3$ will contact extensively and the hetero-architectures are well-formed between them. The XPS measurements are carried out to confirm the coupling mode among atoms in the CSs-BOI-1 and CSs-BOI-5 nanojunctions, as shown in Fig. 3.10. In the high-resolution spectra (Fig. 3.10b), two strong peaks at around 158.4 and 163.8 eV can be assigned to $Bi4f_{7/2}$ and $Bi\ 4f_{5/2}$, which are the characteristic peaks of Bi^{3+} in $BiOIO_3$. The peaks observed including 624.8 eV ($I\ 3d_{5/2}$) and 636.0 eV ($I\ 3d_{3/2}$) can be attributed to $BiOIO_3$ and it indicates that the state of I in the sample is +5 valence. In addition, it can be observed that a new peak appearing at ca. 620.4 eV for sample CSs-BOI-5 is the characteristic peak of BiOI and it indicates that I^- exists in the sample [19], which confirms the results revealed by XRD analysis and TEM observation. In the CSs-BOI-1, the O 1s core level spectra can be fitted by three peaks at binding energies of around 530.4, 532.6, and 534.3 eV (Fig. 3.10d). The peak at 530.4 eV is the characteristic peak of Bi-O linkage in $BiOIO_3$, and the other two peaks at around 532.2 and 534.3 eV can be attributed to the C–O (and O–H), and C–OH bonds respectively [20], which move to 531.0, 532.6 and 534.2 eV after the content of CSs rises to 5%. Meanwhile, carbon is detected by XPS with C 1s binding energy as shown in Fig. 3.10e. It indicates that the main C 1s peak is dominated by elemental carbon at 284.0 eV, which is mainly attributed to the adventitious element of C from the C tape used during the preparation of the sample [21], and the peak at 285.6 is the characteristic peak of the combination of CAC bands, which belong to CSs [22], while the peak at 288.1 eV can be attributed to the C–O [23]. There is some shift of C 1s binding energy when the content of CSs rises to 5%. Such inner shift of the orbits originates from the interaction of $BiOIO_3$ with CSs. In XPS, not only the information on the binding energy of a specific element can be obtained, but also the total density of states (DOS) of the valence band (VB) can be attained. Some new localized electronic states below 2.0 eV above the valence band edge are observed for the CSs-BOI-5 sample compared with pure $BiOIO_3$ sample. These states can be attributed to the C2p orbitals in the doped $BiOIO_3$ [24, 25]. It indicates that the doped carbon does not change the valence band position but creates some isolated localized states above the valence band edge, and these localized states are directly responsible

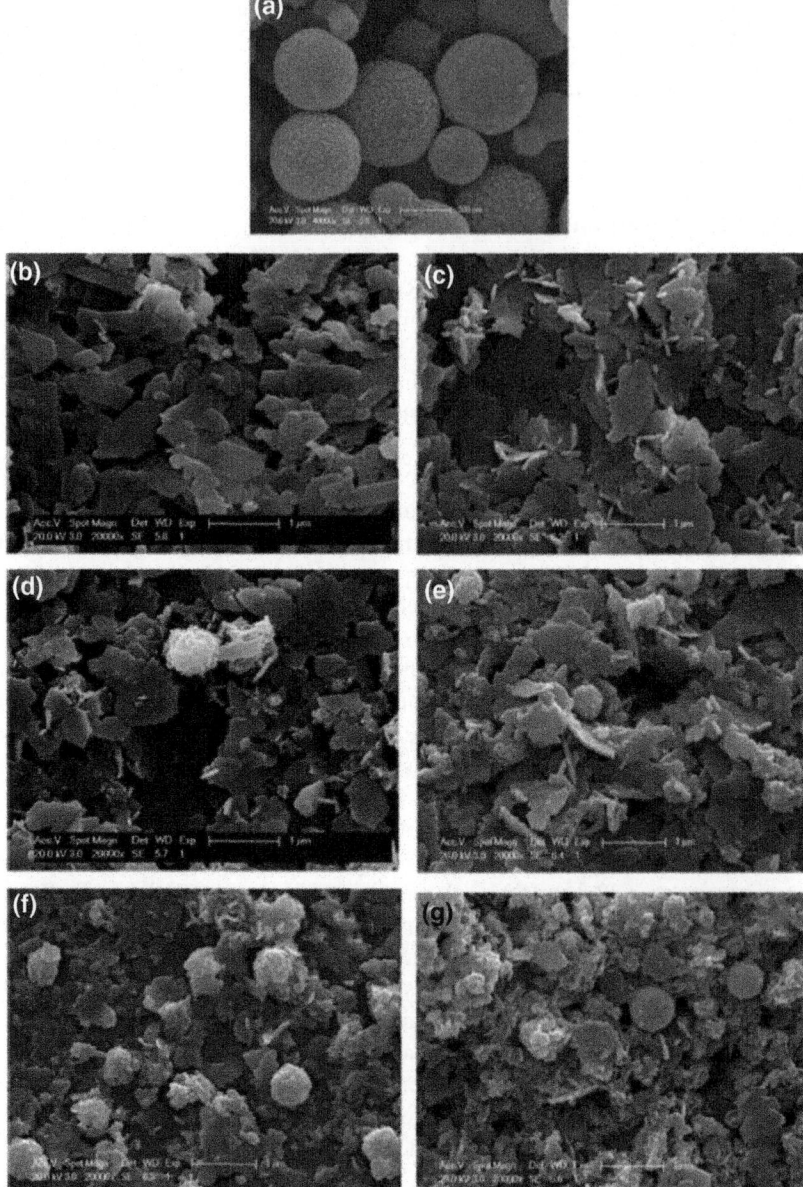

Fig. 3.9 SEM image of CSs (**a**), P-BiOIO$_3$ (**b**), CSs-BOI-0.5 (**c**), CSs-BOI-1 (**d**), CSs-BOI-3 (**e**), CSs-BOI-5 (**f**), CSs-BOI-10 (**g**). Reprinted from Ref. [18], Copyright 2016, with permission from Elsevier

for the electronic origin of band gap narrowing and visible light photoactivity of CSs-BiOI/BiOIO$_3$ [25, 26]. Figure 3.12a shows the FTIR spectra of CSs, P-BiOIO$_3$ and

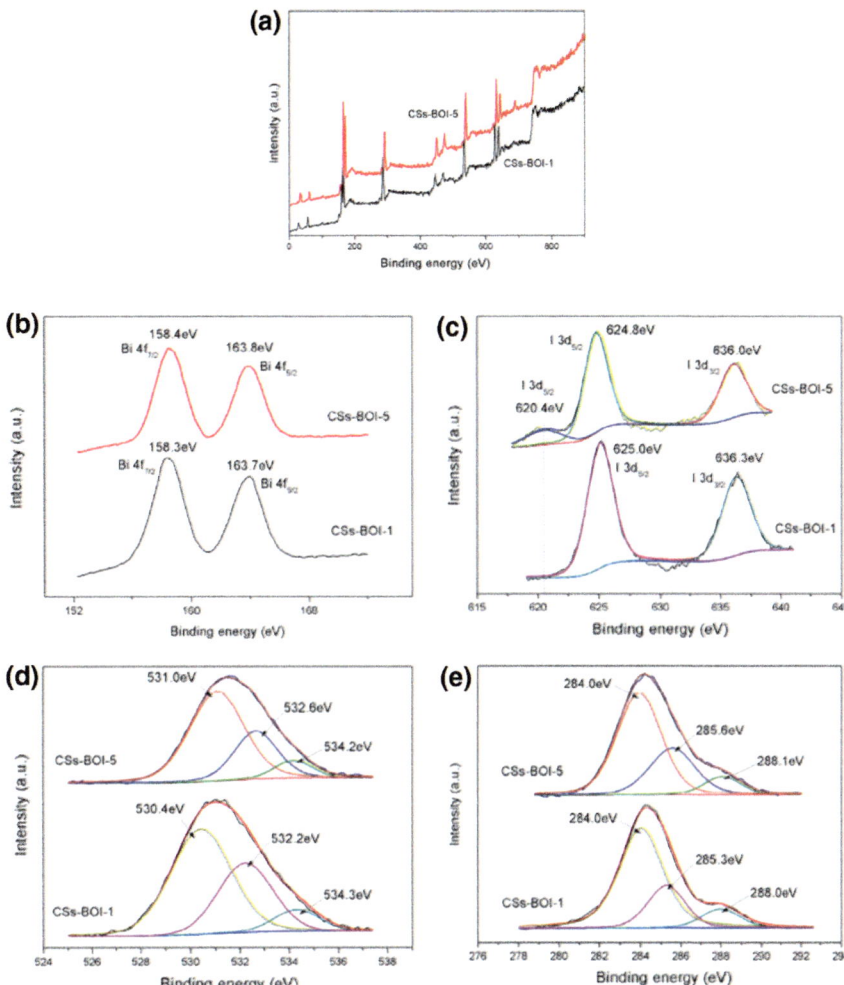

Fig. 3.10 XPS spectra of CSs-BOI-1 and CSs-BOI-5 samples: **a** survey, **b** Bi 4f, **c** I 3d, **d** O 1s and **e** C 1s. Reprinted from Ref. [18], Copyright 2016, with permission from Elsevier

CSs-BOI-1. For P-BiOIO$_3$ and CSs-BOI-1, two peaks observed at about 687 cm^{-1} and 773 cm^{-1} can be attributed to vibration of I-O bond. Besides, the absorption peaks at around 520 cm^{-1} are assigned to the vibration of Bi-O bond, and a mild red shift can be discovered with CSs doping in Fig. 3.12b. It can be attributed to the doped carbon atoms reacting with BiOIO$_3$ to produce BiOI, which gets the Bi–O bond at 574 cm^{-1}. At the same time, the I–O bond and Bi–O bond decreased gradually with CSs content increasing. Moreover, the bending vibrations (3435 cm^{-1}) corresponding to the O–H vibration include hydroxyl groups and molecular water, increasing from 0.5 to 10%. This can be because that the nanoparticles become smaller, which is based

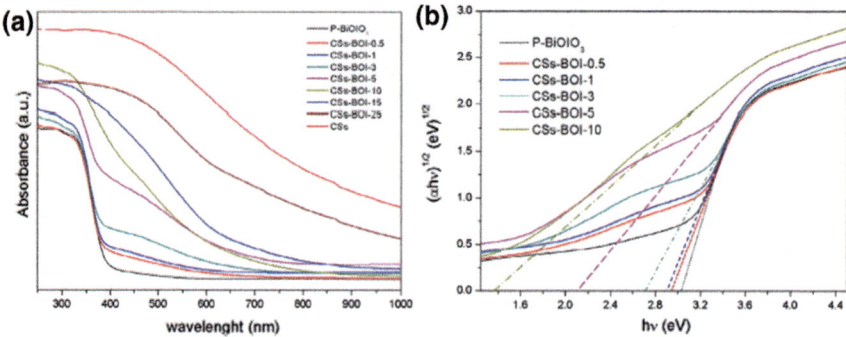

Fig. 3.11 UV-vis diffuse reflectance spectrum (**a**) and the associated (ahm) 1/2 versus (hm) plot (**b**) of the CSs, BiOIO$_3$ and CSs-BiOI/BiOIO$_3$. Reprinted from Ref. [18], Copyright 2016, with permission from Elsevier

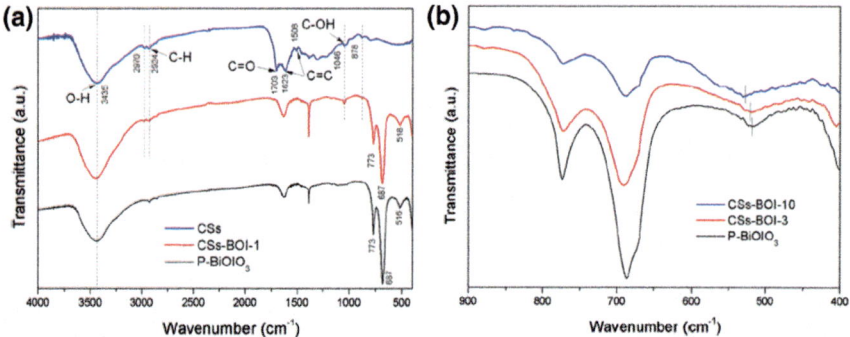

Fig. 3.12 FTIR spectra of CSs, P-BiOIO$_3$ and CSs-BOI-1 composites (**a**) and enlarged view of the spectra region between 400 and 900 cm^{-1} (**b**). Reprinted from Ref. [18], Copyright 2016, with permission from Elsevier

on the SEM observation shown in Fig. 3.9. With the CSs doping increasing, more and more BiOI is produced and clogging the pores, so the SBET generally decreased. Nevertheless, the SBET of CSs/BiOIO$_3$ were much more than that of single CSs and BiOIO$_3$. In general, the larger surface area is conducive to the photocatalytic activity.

The optical properties of all samples have been tested by UV-vis diffuse reflectance spectroscopy, shown in Fig. 3.11. As expected, the pure BiOIO$_3$ displays intense absorption in the UV light region with fundamental absorption edge below 400 nm. CSs powder, with the color of black, exhibits strong absorption in whole range of wavelength employed (200–800 nm). With the increase of CSs content in the composite catalysts, the optical absorption between 400 and 550 nm of CSs-BOI samples is also enhanced, while the absorption edge displays the red shift step by step, which is in agreement with the observed color change from white to claybank. Due to induced interphase interaction between BiOIO$_3$, BiOI and CSs, the absorption range extends, which is in agreement with other papers. The long-tail absorption in the visible-light region can be attributed to electronic interactions between carbon species, the produced BiOI and $BiOIO_3$. On one hand, the carbon atoms are incorporated into the interstitial positions of $BiOIO_3$ lattice, leading to several localized occupied states in the gap, and intact CSs coating on the surface of $BiOIO_3$ may be considered as photosensitizer, which are likely to carry out a charge transfer process and responsible for the photosensitized photocatalysis [20, 22]. On the other hand, pure BiOI displays strong photo-absorption from visible light shorter than 700 nm, which is combined with $BiOIO_3$ and narrows the band gap of $BiOIO_3$.

References

1. J.H. Zeng, B.B. Jin, Y.F. Wang, Facet enhanced photocatalytic effect with uniform single-crystalline zinc oxide nanodisks. Chem. Phys. Lett. **472**, 90–95 (2009)
2. J. Becker, K.R. Raghupathi, J. St Pierre, D. Zhao, R.T. Koodali, Tuning of the crystallite and particle sizes of ZnO nanocrystalline materials in solvothermal synthesis and their photocatalytic activity for dye degradation. J. Phys. Chem. C **115**, 13844–13850 (2011)
3. F. Xu, Y. Shen, L. Sun, H. Zeng, Y. Lu, Enhanced photocatalytic activity ofhierarchical ZnO nanoplate-nanowire architecture as environmentally safe and acilely recyclable photocatalyst. Nanoscale **3**, 5020–5025 (2011)
4. A. Umar, M.S. Akhtar, A. Al-Hajry, M.S. Al-Assiri, G.N. Dar, M.S. Islam, Enhancedphotocatalytic degradation of harmful dye and phenyl hydrazine chemicalsensing using ZnO nanourchins. Chem. Eng. J. **262**, 588–596 (2015)
5. M.K. Kavitha, S.C. Pillai, P. Gopinath, H. John, Hydrothermal synthesis of ZnO decorated reduced graphene oxide: understanding the mechanism of photocatalysis. J. Environ. Chem. Eng. **3**, 1194–1199 (2015)
6. C.J. Xu, Y.Z. Wang, H.Y. Chen, Rapid and simple synthesis of 3D ZnO microflowers at room temperature. Mater. Lett. **158**, 347–350 (2015)
7. E. Pál, V. Hornok, A. Oszkó, I. Dékány, Hydrothermal synthesis of prism-like and flower-like ZnO and indium-doped ZnO structures. Colloids Surf. A **340**, 1–9 (2009)
8. X. Zhao, F. Lou, M. Li, X. Lou, Z. Li, J. Zhou, Sol-gel-based hydrothermal method for the synthesis of 3D flower-like ZnO microstructures composed of nanosheets for photocatalytic applications. Ceram. Int. **40**, 5507–5514 (2014)
9. J.L. Yang, S.J. An, W.I. Park, G.C. Yi, W. Choi, Photocatalysis using ZnO thin films and nanoneedles grown by metal-organic chemical vapor deposition. Adv. Mater. **16**, 1661–1664 (2004)
10. M. Parhizkar, M. Kumar, P.K. Nayak, S. Singh, S.S. Talwar, S.S. Major, R.S. Srinivasa, Nanocrystalline ZnO films prepared by pyrolysis of Zn-arachidate LB multilayers. Colloids Surf. A **257–58**, 445–449 (2005)
11. Ruixing Zhou, Wu Jiang, Jing Zhang, Huan Tian, Pankun Liang, Tao Zeng, Lu Ping, Jianxing Ren, Tianfang Huang, Xiao Zhou, Pengfei Sheng, Photocatalytic oxidation of gas-phase Hg^0 on the exposed reactive facets of $BiOI/BiOIO_3$ heterostructures. Appl. Catal. B **204**, 465–474 (2017)
12. H.W. Huang, K. Xiao, K. Liu, S.X. Yu, Y.H. Zhang, In situ composition-transforming fabrication of BiOI/BiOIO3 heterostructure: semiconductor p-n junction and dominantly exposed reactive facets. Cryst. Growth Des. **16**, 221–228 (2016)
13. H. Cheng, B. Huang, Y. Dai, X. Qin, X. Zhang, One-step synthesis of the nanostructured AgI/BiOI composites with highly enhanced visible-light photocatalytic performances. Langmuir **26**, 6618–6624 (2010)
14. Y. Xu, M.A. Schoonen, The absolute energy positions of conduction and valence bands of selected semiconducting minerals. Am. Mineral. **85**, 543–556 (2000)

15. M.C. Long, W.M. Cai, J. Cai et al., Efficient photocatalytic degradation of phenol over Co_3O_4/$BiVO_4$ composite under visible light irradiation. J. Phys. Chem. B. **110**(41), 20211–20216 (2006)
16. H. Zhang, R. Zong, J. Zhao et al., Dramatic visible photocatalytic degradation performances due to synergetic effect of titanium dioxide with PANI. Environ. Sci. Technol. **42**(10), 3803–3807 (2008)
17. F. Dong, T. Xiong, Y.J. Sun, Y.X. Zhang, Y. Zhou, Controlling interfacial contact and exposed facets for enhancing photocatalysis via 2D-2D heterostructure. Chem. Commun. **10**, 1039 (2010)
18. J. Wu, X. Chen, C. Li, Y. Qi, X. Qi, J. Ren, B. Yuan, B. Ni, R. Zhou, J. Zhang, T. Huang, Hydrothermal synthesis of carbon spheres—BiOI/$BiOIO_3$ heterojunctions for photocatalytic removal of gaseous Hg^0 under visible light. Chem. Eng. J. **304**, 533–543 (2016)
19. G. Dai, J. Yu, G. Liu, Synthesis and enhanced visible-light photoelectrocatalytic activity of p-n junction BiOI/titanium dioxide nanotube arrays. J. Phys. Chem. C **115**, 7339–7346 (2011)
20. J. Zhong, F. Chen, J. Zhang, Carbon-deposited titanium dioxide: synthesis, characterization, and visible photocatalytic performance. J. Phys. Chem. C **114**, 933–939 (2010)
21. Y. Zhang, Z. Zhao, J. Chen, L. Cheng, J. Chang, W. Sheng, C. Hu, S. Cao, C-doped hollow titanium dioxide spheres: in situ synthesis, controlled shell thickness, and superior visible-light photocatalytic activity. Appl. Catal. B **165**, 715–722 (2015)
22. H. Li, D. Wang, H. Fan, P. Wang, T. Jiang, T. Xie, Synthesis of highly efficient C doped titanium dioxide photocatalyst and its photo-generated charge-transfer properties. J. Colloid Interface Sci. **354**, 175–180 (2011)
23. J. Zhuang, Q. Tian, H. Zhou, Q. Liu, P. Liu, H. Zhong, Hierarchical porous titanium dioxide@C hollow microspheres: one-pot synthesis and enhanced visible-light photocatalysis. J. Mater. Chem. **22**, 7036–7042 (2012)
24. X. Chen, P.-A. Glans, X. Qiu, S. Dayal, W.D. Jennings, K.E. Smith, C. Burda, J. Guo, X-ray spectroscopic study of the electronic structure of visible-light responsive N, C and S doped titanium dioxide. J. Electron. Spectrosc. **162**, 67–73 (2008)
25. X. Chen, C. Burda, The electronic origin of the visible-light absorption properties of C, N and S doped titanium dioxide nanomaterials. J. Am. Chem. Soc. **130**, 5018–5019 (2008)
26. S. Sakthivel, H. Kisch, Daylight photocatalysis by carbon-modified titanium dioxide. Angew. Chem. Int. Ed. **42**, 4908–4911 (2003)

Chapter 4
Modified Photocatalysts

Abstract For photocatalyst, changing its appearance and doping with other materials are the effective methods to enhance its photocatalytic activity. For titanium-based photocatalysts, we studied the titanium dioxide hollow microspheres and anatase titanium dioxide with co-exposed (001) and (101) planes. The results show that controlling of its morphology can effectively improve the photoactivity. Meanwhile, doping titanium dioxide with metal oxide and nonmetal (CuO/titanium dioxide, V_2O_5/titanium dioxide, carbon spheres supported CuO/titanium dioxide, carbon decorated In_2O_3/titanium dioxide) also can improve the separation efficiency of photogenerated electrons and holes. Moreover, the as-prepared materials were characterized by XRD, XPS, and TEM to study its physical and chemical properties. In addition to titanium-based photocatalysts, we have also researched zinc-base photocatalysts. Zinc-base photocatalysts show enhanced photocatalytic performance through combining with metals and nonmetals. Bismuth-based photocatalyst is a hot spot of research in recent years, and we present the doped bismuth-based photocatalysts modified by metal or nonmetal. Research shows that modified bismuth-based photocatalysts with metal or nonmetal is an efficient method to let its band gap narrow. Graphene has attracted the attention of scientist for its excellent performance, and we have studied the graphene supported titanium dioxide photocatalysts and its physical and chemical properties.

Keywords Hollow microspheres · Metal modified titanium dioxide Nonmetal modified titanium dioxide

4.1 Morphology Controlled Photocatalyst Synthesis Methods

4.1.1 Titanium Dioxide Hollow Microspheres Photocatalysts

Due to its low density, large specific surface area and good photocatalytic activity, titanium dioxide hollow structural materials have attracted more and more attention.

Titanium dioxide hollow structured materials have been successfully prepared and applied in many fields such as the organic pollutants decomposition [1]. Titanium-dioxide hollow spheres were produced template-free with hydrothermal method by using trifluoroacetic acid (TFA) and $Ti(SO_4)_2$ as raw materials [2]. The photocatalytic performance of titanium dioxide hollow spheres for the Hg^0 photocatalytic oxidation removal from coal-fired flue gas was studied.

4.1.2 Anatase Titanium Dioxide with Co-exposed (001) and (101) Facets

In addition, some studies present that the microstructure of surfaces and interfaces in metal oxide nanoparticles can be controlled by the chemical parameters of synthesis process, which affect their chemical and physical properties [3, 4]. Some studies show that the synergistic effect of different planes is a probable mechanism for enhancing photoactivity [5].

Herein, we demonstrate anatase titanium dioxide with co-exposed (001) and (101) planes and search the ideal ratio of these planes to achieve the best photocatalytic performance. The formation of surface heterojunctions between the (001) and (101) planes facilitates the separation of photoinduced electron-hole pairs [6]. Moreover, the as-prepared materials are used for removal of elemental mercury (Hg^0), which is difficult to be removed by photocatalysis owing to high oxidation-reduction potentials. At the same time, the possible reaction mechanism is set forth based on the characterization and experimental results.

4.2 Metal or Nonmetal Modified Zinc Base Photocatalysts

Doping metal or nonmetal atoms can change the photoelectric properties of ZnO to expand its spectral response to the visible light area since it can effectively reduce the band gap of semiconductors [7]. The doped metal/nonmetal atoms alter the coordination environment of Zn atom and adjust the electronic structure of ZnO by adding localized electronic energy levels in the band gap (Fig. 4.1). The dopant energy level is temporally lied under the CB, where photogenerated charge carriers are trapped and photocatalytic activities are thus increased [8–10].

4.2.1 Doping Metals

Doping metal atoms, such as alkali metals, rare earth metals, transition metals, and noble metals, to ZnO crystal lattice has been widely studied [11]. Metals doped

4.2 Metal or Nonmetal Modified Zinc Base Photocatalysts

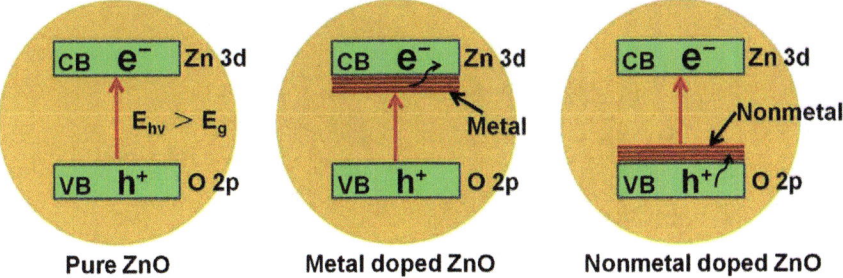

Fig. 4.1 Schematic of the comparison of the band structures of pure, metal-doped ZnO, and nonmetal-doped ZnO. Reprinted from Ref. [3], copyright 2018, with permission from Elsevier

to ZnO likely enlarge the visible light response of ZnO and increase its quantum efficiency [12]. Therefore, doping ZnO with metals may increase the photocatalytic activity of ZnO.

4.2.1.1 Doping Transition Metals

Transition metals and their cations have an unfulfilled d subshell. Transition metals with similar atomic radii to the atomic radius of Zn can be readily doped into the ZnO lattice. Many transition metals, including Fe [13], Co [14], Ni [15], Mn [16], Cr [17], V [18], Cu [19], and Zr [20], have been doped to ZnO to reduce the band gap of this compound [21]. The red shift is engendered by the developing charge move between d electrons of transition metals and the CB or VB of ZnO [22]. Moreover, a metal atom under this condition may produce a new electron state in the band gap of ZnO, which in turn traps excited electrons and restrains the recombination of the e^-/h^+ pair. So, the type and concentration of these doping transition metals are the main factors affecting the photocatalytic performance of ZnO [23].

Xu et al. [24] Synthesized CuO/ZnO nanocomposites by a facile one-step hydrothermal way. Compared with pure ZnO, the photocatalytic performance of the CuO/ZnO nanocomposites is greatly improved, which is attributed to the band coupling between ZnO and CuO and the improvement of solar energy's utilization efficiency. Shown as Fig. 4.2, it is found that the CuO contents in the nanocomposite have a great influence on the photocatalytic performance. When the molar ratio of Zn:Cu is 2:1 in the growth solution, the ensuing CuO/ZnO nanocomposite presents the best photocatalytic performance. This indicates that CuO/ZnO nanocomposite is a coupling system for the application as a photocatalyst. Only when each unit in this system fully exerts its function and a synergistic effect is formed, the nanocomposite can achieve the best photocatalytic performance.

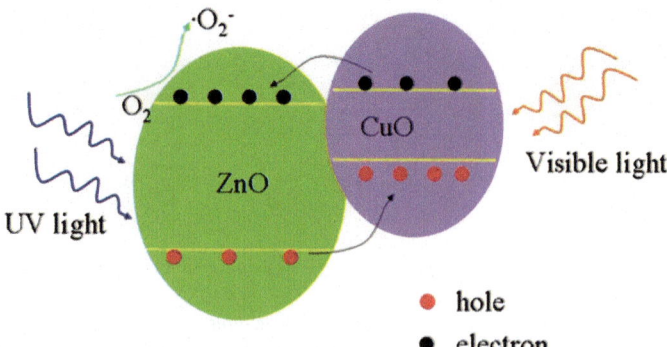

Fig. 4.2 Schematic diagram of excitation and transfer of electrons and holes for ZnO/CuO heterojunction under irradiation of light. Reprinted from Ref. [24], copyright 2018, with permission from Elsevier

4.2.1.2 Doping Rare Earth Metals

Rare earth (RE) metals are good doping elements, which can be used for altering the electronic structure of ZnO and increasing visible light absorption. Doping RE metals forms a localized impurity level in the band structure of ZnO and changes the band of ZnO. The electronic structure of ZnO is also influenced by the charge transfer between the VB or CB of ZnO and the 4f or 5d electrons of RE metals [25–27]. Though RE doping has some merit, this tech is restricted by the low doping saturation of RE ions in the ZnO crystal lattice due to the aniso in ionic radii and a small mismatch between the energy level positions of RE ions and ZnO [28].

Thi and Lee [29] successfully synthesized a chain of La-doped ZnO photocatalysts with various ratios of La dopant (0.5, 1.0 and 1.5 wt%) through a simple fabrication process. La^{3+} was doped into the ZnO lattice and the band gap of ZnO reduced from 3.15 eV to 3.02, 2.94 and 3.09 eV with the use of 0.5, 1.0 and 1.5 wt% La-doped ZnO respectively. 1.0 wt% La-doped ZnO presents the maximum photocatalytic degradation performance (99 and 85% of removing efficiency on paracetamol and TOC, separately) after 3 h exposure to the visible light irradiation, while very low photocatalytic activity was watched for pure ZnO. HPLC and GC-MS analysis can identify many intermediates and/or by-products of APAP photocatalytic decomposition. According to determined chemical structure, a photocatalytic degradation reaction mechanism of APAP in aqueous solution was proposed in detail, including possible forms of radical generation and following reaction pathways, as shown in Fig. 4.3.

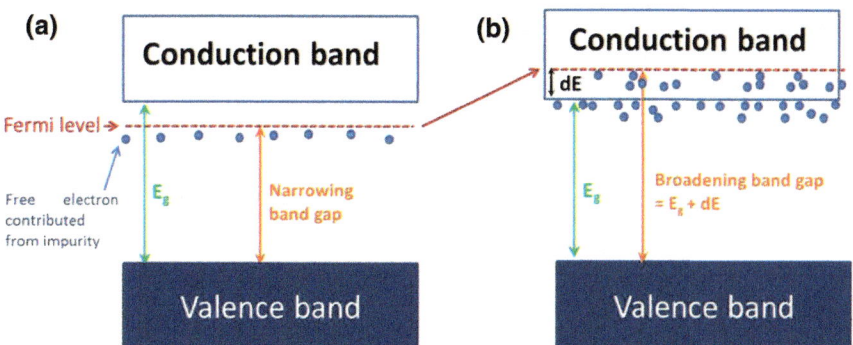

Fig. 4.3 Proposed mechanism for **a** narrowing band gap of 0.5–1.0 wt% La-doped ZnO and **b** broadening band gap of 1.5 wt% La-doped ZnO. Reprinted from Ref. [29], copyright 2018, with permission from Elsevier

4.2.1.3 Doping Alkali Metals

Alkali metal-doped ZnO can improve the photocatalytic activities for organic pollutant photodegradation. For instance, the photocatalytic activity of Na-doped ZnO is higher than that of pure ZnO in dye photodegradation [30, 31]. Photodegradation efficiency is also influenced by the doping amount of alkali metals. For example, in the degradation process of RhB, the photocatalytic performance of low concentration K-doped ZnO is very high [32]. Different types of doping elements cause various effects on photocatalytic activities. The photocatalytic activity of Li-doped ZnO in 4-NP degradation is higher than that of Na- or K-doped ZnO because Li^+ can capture excited electrons [33, 34]. The visible-light photocatalytic activity of Mg-doped ZnO in MB degradation is increased due to the narrowing of band gap and efficient charge carrier transfer [35]. The widening of band gap caused by Mg^{2+} doping is ascribed to Mosse Burstein effect resulted from electrons trapped in oxygen vacancies. Due to the differences in the ionic radii and electronegativities between Zn^{2+} and Mg^{2+}, Mg^{2+} doping enhanced the ability of oxygen vacancies to trap more excited electrons and thus increase electron concentrations. But, doped Mg^{2+} can serve either as the lattice substitution of Zn^{2+} or the interstitial occupation entering the ZnO lattice which also influences photocatalytic performance [36].

4.2.2 Doping Nonmetals

It has been widely observed that the replacement of lattice oxygen atoms can be accomplished by doping nonmetals to the ZnO lattice [37, 38]. Doping elements should not only require less electronegativity than oxygen, it should also have a similar atomic radius as the O atom, and both conditions are important factors for effective doping [39]. Dopants, such as C, N, and S, can form straight energy levels in the band gap and thus enhance visible-light photocatalytic activities [40, 41].

Yu et al. [42] estimated the electronic structures, optical properties and effective masses of charge carriers of pure, N-, C- and S-doped ZnO based on plane-wave pseudo-potential method of DFT calculation (Figs. 4.4 and 4.5). The electronic structures present that the valence band of pure ZnO is mainly composed of Zn 3d states (lower) and O 2p states (upper), as the conduction band is equally populated by Zn 4s and 3p states. Both N and C doping have two major roles: (1) enhancing the Fermi level electron density and (2) pullining vacant (impurity) states into the band gap, mostly because of their nature of p-type doping. The introduced impurity states by N and C doping can promote the excitation of electrons, futhermore C doping can also narrow the gap between the Fermi level and conduction band. Without the appearance of vacant states, the increased photocatalytic activity of S-doped ZnO is attributed to direct band gap narrowing. The calculation of optical properties show that N and C doping can induce much stronger light absorption of visible and ultraviolet light. It is worth noting that C-doped ZnO shows the most intense light absorption over the whole spectrum, resulting from its most significantly electronically deficient character. The calculated effective masses of charge carriers demonstrate that as an n-type semiconductor, ZnO contains light electrons and heavy holes. When doped with N, C or S, the photogenerated electrons and holes all become lighter and heavier, separately. The recombination rates of photogenerated electron-hole pairs are evaluated as C doping «N doping < S doping < pure ZnO. These theoretical studies may provide new insights into the ground understanding of the fundamental mechanism for the improved photocatalytic activity of N-, C- and S-doped ZnO.

4.3 Metal or Nonmetal Modified Titanium Dioxide Photocatalysts

4.3.1 CuO/Titanium Dioxide Photocatalysts

Due to its nontoxicity, low cost, and high chemical stability, titanium dioxide is one of the most widely accepted semiconductor photocatalysts for oxidation of elemental Hg. Yet, titanium dioxide can only be activated by ultraviolet (UV) irradiation, which accounts for only about 4% of the total solar radiation. Moreover, low rate of electron transfer and high rate of recombination between excited electron-hole pairs often caused low quantum yield rate and a limited photooxidation rate [43, 44]. Many researchers tried many methods to modify the photocatalyst to make photocatalysis under visible light more practical and enhance its photocatalytic oxidation efficiency. These methods include surface modification [1, 45], doping with metal or nonmetal ion [46, 47], combining with other semiconductors or metal oxides [48, 49], and etc. In these methods, doping metal oxides into titanium dioxide is an appealing method to stop the recombination of charge carriers and reduce the band gap. Therefore the

4.3 Metal or Nonmetal Modified Titanium Dioxide Photocatalysts

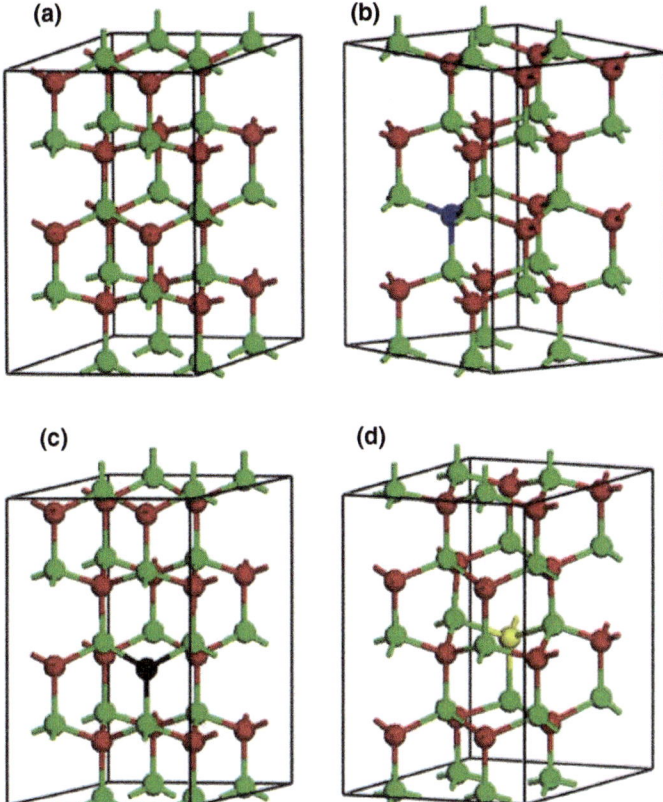

Fig. 4.4 Schematic illustration of the optimized crystal structures of $2 \times 2 \times 2$ supercells for pure **a** N-doped **b** C-doped **c** S-doped **d** ZnO; the green and red balls represent zinc and oxygen atoms, respectively; the blue (**b**), black (**c**), yellow (**d**) balls stand for the doped nitrogen, carbon and sulfur atoms, respectively (For interpretation of the references to color in this figure legend, the reader is referred to the web version of the article). Reprinted from Ref. [42], copyright 2018, with permission from Elsevier

photocatalytic efficiency could be enhanced and the response range could expand to visible light region. CuO is one of the low-cost transition metal oxides. CuO has been considered as a good active component for reducing water under sacrificial condition [50, 51], owing to the fact that it can stop the recombination between photogenerated electrons and holes. Meanwhile, it can reduce the energy of the band gap. Yet, few researches have reported the use of CuO catalyst supported on titanium dioxide as a system for Hg^0 photocatalytic oxidation. In addition, few studies reported the influence of different light sources on the photocatalytic oxidation.

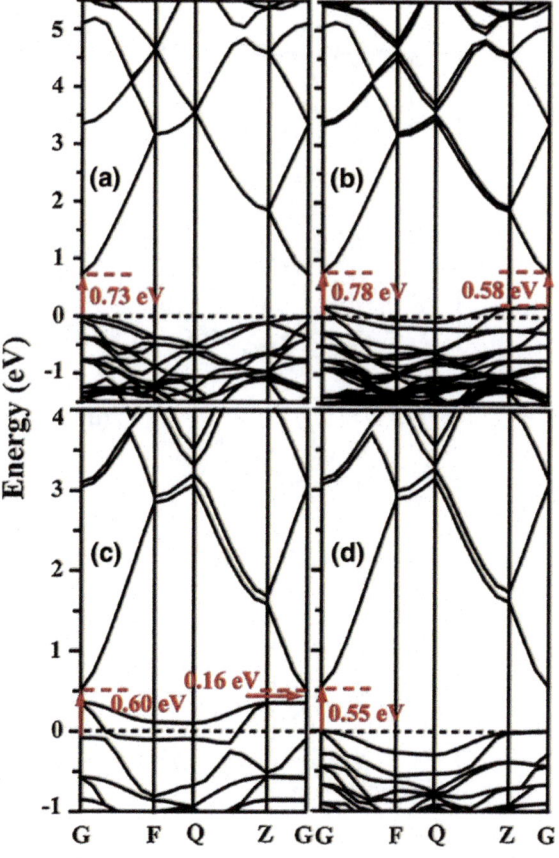

Fig. 4.5 Comparison of band structures and possible excitation energy of pure **a** N-doped **b** C-doped **c** and S-doped **d** ZnO. Reprinted from Ref. [42], copyright 2018, with permission from Elsevier

4.3.2 V_2O_5/Titanium Dioxide Photocatalysts

Compared with other means, the combination of metal oxides with titanium dioxide support is an efficient method to repress the recombination of photogenerated electrons and holes and reduce the band gap [52, 53].

As an important metal oxide photocatalyst with a narrow band gap (about 2.24 eV), V_2O_5 has been compounded with titanium dioxide host by multi-methods and presented visible light-induced photocatalytic activity [54]. In addition, V_2O_5 is beneficial to the separation of photogenerated electron-hole pairs, the expansion of the absorption spectrum and the increase of the surface charge carrier transfer rate [55]. At the same time, introduction of V_2O_5 to the titanium dioxide support can produce heterojunctions on the interface. Lately, various of heterojunction semiconductors have been progressively extended to the field of photocatalysis [56–61]. Xie et al. [62] prepared V_2O_5/titanium dioxide core-shell spherical and solid sphere nanostructures showing an increased photocatalytic oxidation of arsenite. Ren et al. [63] found that V-doped titanium dioxide using the nanotubular titanic acid as titanium precursor presents significant photocatalytic activity for propylene's degradation. Wu et al. [64] tested the photoactivity of C-doped V_2O_5/titanium dioxide, presenting improved performance in degradation of gas-phase toluene. Wang et al. [65] reported that 1D V_2O_5/titanium dioxide heterostructure demonstrated increased photocatalytic performance towards degradation of Rhodamine B. Thus, the heterojunction between V_2O_5 and titanium dioxide can improve the photocatalytic performance of host by boosting the separation and migration of photogenerated electron-hole pairs at the interfaces. At the same time, oxygen vacancies forming during the preparation process can further promote the separation of photoexcited electrons and holes [66]. Yet, there are few reports about one photocatalyst with homo-hetero junctions and oxygen vacancies.

We made a V_2O_5/rutile-anatase (vanadium oxide loaded on titanium oxide) photocatalyst system with oxygen vacancy, by facile incipient wet impregnation method. Via combining both homo- and hetero-junctions into one photocatalyst with oxygen vacancy, the photogenerated electron-hole pairs reach a more effective separation and migration due to the synergetic effects of homo-hetero junctions and oxygen vacancy, which can improve the photocatalytic efficiency. Meanwhile, the V_2O_5/rutile-anatase samples were used for solving the environmental contaminations to examine its photocatalytic performance.

4.3.3 Carbon Spheres Supported CuO/Titanium Dioxide Photocatalysts

Many researches have been carried out to modify the photocatalyst to make photocatalysis under visible light more practical and enhance photocatalytic oxidation efficiency. These studies include surface modification [1, 45], metal or metal oxides doping [67, 68], anionic nonmetal doping [47, 69], and etc. The metal doping of titanium dioxide often has poor thermal stability, photo-corrosion [70]. On the contrary, it has been reported that, non-metal (C, N, S, etc.) doping exhibits a red-shifted absorption edge and extends the photocatalytic activity into the visible light region for titanium dioxide. Particularly, carbon doping exhibits sizable possible advantages over other types of non-metal doping, which may be related to the following reason: carbon not only can act as adsorbent or support, but also can act as sensitizer and transfer electrons to the semiconductors, triggering the formation of very reactive radicals to enhance photocatalytic activity of semiconductor for target reactions [71]. Copper oxide (CuO), a p-type semiconductor, has been broadly used for the reduction of water under sacrificial condition [72–77] since it can stop the recombination between photogenerated electrons and holes. Simutaneously it can reduce the energy of the band gap. Yousef et al. [78] studied the CuO nanoparticles (NPs)-doped titanium dioxide nanofibers for three azo dyes photodegradation under visible light. Jia et al. [79] synthesized graphite/C-doped titanium dioxide composite, presenting a significant improved photocatalytic performance for degradation of MO. While, there have been few studies reporting the use of CuO catalyst and carbon supported on titanium dioxide as a system for Hg^0 photocatalytic oxidation.

4.3.4 Carbon Decorated In_2O_3/Titanium Dioxide Photocatalysts

Titanium dioxide (TiO_2), one of the most famous semiconductor photocatalysts, is with strong oxidative property, non-toxicity, and high chemical stability, and it has been observed to work out environmental pollutant and energy crisis [80]. However, its large band gap (3.02 eV for rutile, 3.20 eV for anatase) hinders the application of titanium dioxide photocatalysts, which can be excited in the ultraviolet light [81]. In addition, photoexcited electrons and holes are easier recombined and difficult to migrate, resulting in a limited photooxidation rate. In order to improve photocatalytic activity of titanium dioxide carriers, many researches have been carried out to modify it to enhance its optical properties and extend the absorption edge to visible light. These studies include surface modification [82], incorporating metal or nonmetal ion [83, 84], and combining with metal oxides [85].

Combining metal oxides with titanium dioxide host is a significant method to promote the migration of photoexcited electrons and holes and reduce its band gap [86]. Indium oxide (In_2O_3), a vital n-type semiconductor, which indirect band gap is nearly 2.6 eV, has been used to reduce the band gap of host [87, 88]. At the same time, the formation of In_2O_3/titanium dioxide heterojunction can efficiently improve the interfacial charge transfer and the separation of the photoexcited electron-hole pairs, and thus enhance the photoactivity. In addition, the conduction band (CB) levels of combined semiconductors in the proper place can improve the migration of photoexcited electrons [89]. The band gap of In_2O_3 is lower than that of titanium dioxide, but the CB of In_2O_3 (E_{CB} for $In_2O_3 = -0.63$ V versus NHE, which means the CB potential can be determined as the x-intercept of Mott-Schottky curve) is higher than that of titanium dioxide (E_{CB} for titanium dioxide $= -0.4$ V vs. NHE). Thus, the formed heterostructure is an effective junction to improve the separation of photoinduced electron-hole pairs when coupling In_2O_3 and titanium dioxide together. Although combining metal oxides with titanium dioxide support can promote its activity, the introduced additional metallic ions also act as a recombination center, which leads to excited electrons and holes to be combined once again. Doping non-metal with titanium dioxide support, such as carbon, nitrogen and sulfur, is an effective measure to expand the absorbing edge to the visible light region for titanium dioxide and suppress recombination of the excited electrons and holes [90, 91]. In particularly, doping carbon with host presents considerable potential superiorities compared with other genres of non-metal doping [92]. Carbon not only can act as adsorbent or support, but also can act as sensitizer, which can improve electrons migrating to the semiconductors [93]. So, it is advantageous to use the synergistic functions of metal oxides and non-metal for expanding the absorbing edge and improving the separation of excited electrons and holes. It is an effective method to solve environmental pollutant and energy crisis by using In_2O_3 and carbon co-doped titanium dioxide support as a photocatalyst system.

In our work, a various of mole percentage of indium and stationary carbon co-modified titanium dioxide ternary materials were prepared by a facile incipient wet impregnation means. Doping titanium dioxide with carbon can produce an additional carbon-doping level above the valence band (VB) of titanium dioxide, simultaneously, an effective heterojunction between In_2O_3 and titanium dioxide is formed too. In this ternary nanocomposite, band gap is reduced, and photoinduced electrons and holes are effectively separated.

4.4 Metal or Nonmetal Modified BiVO$_4$ Photocatalysts

Metal doping inhibits photoelectron-hole interaction by introducing defect positions or changing crystallinity into the lattice. Chala et al. synthesized Fe loaded $BiVO_4$ samples by hydrothermal method. Under visible light irradiation, when Fe reached the optimal loading of 5.0%, the photocatalyst had the best photodegradation performance for methylene blue, with the degradation rate of 81% [94, 95]. Xu et al. doped rare earth elements such as holmium, samarium, ytterbium, europium, gadolinium, neodymium, cerium and lanthanum for $BiVO_4$ photocatalyst. XRD, SEM and XPS results show that the rare earth element ions exist on the surface of the sample in the form of oxides [96, 97]. Non-metallic element doping increases the value band of the semiconductor and reduces the band gap of the semiconductor, thus enhancing the absorption performance of the catalyst to visible light. In addition, non-metallic ion doping can also cause certain defects in the crystal lattice of semiconductors, which can effectively capture the photo-generated electrons and promote the separation of photo-generated charge. Li et al. synthesized F-doped spherical $BiVO_4$ by simple two-step hydrothermal method. Under the irradiation of visible light, the photocatalytic activity of F-doped $BiVO_4$ was higher than that of undoped $BiVO_4$ [97, 98].

4.5 Graphene Supported Titanium Dioxide Photocatalysts

To date, the most widely researched photocatalytic material is titanium dioxide, which has been extensively used as a standard photocatalyst because of its long-term thermodynamic stability, relatively low toxicity, excellent photocatalytic performance and low cost compared to other semiconductor materials. [99–101] However, bare titanium dioxide usage still presents several limitations in practical applications: (i) photogenerated electron-hole pairs can recombine quickly, which results in a low photooxidation rate, [43, 102–104] and (ii) titanium dioxide can only be excited by ultraviolet (UV) irradiation, which is less than 5% of the total solar radiation, due to its wide band gap. [105, 106] It also has been demonstrated that titanium dioxide photocatalytic activity can be affected by many aspects, including crystal phase, crystalline grain size, surface chemical properties, specific surface area and crystallinity [107]. Thereinto, if the size of the titanium dioxide crystalline grain is small, which is favorable to add the number of reactive sites to improve the performance on surface redox reactions, the photo-generated electron will be easily recombined with the hole. High crystallinity can enhance the catalytic activity of titanium dioxide by improving charge separation efficiency, which is usually needed to further thermal treatment. Moreover, thermal treatment at high temperatures may lead to larger particles, and deteriorate photocatalytic performance. Hence, it is necessary to maintain a balance between the size of the crystalline grain and degree of crystallinity [108, 109].

Graphene, due to its excellent charge carrier mobility, large specific surface area, high transparency and good electrical conductivity [110–112], can be selected to prepare composites with titanium dioxide to offer unique advantages for photocatalytic decontamination of air and water [113]. The composite titanium dioxide/RGO have three properties: the excellent adsorptivity of pollutants, extended light absorption range into the visible region, facile charge transportation and separation to reduce electron/hole pair recombination [114, 115]. Therefore, it is expected to become one of the most promising materials in the next generation of photocatalysts.

So far, three major methods are adopted to synthesize GO, a precursor of RGO. (i) the Brodie method [116], (ii) the Staudenmaier method [117], and (iii) the Hummers method. We researched a simple method to prepare different oxidation levels of GO without careful temperature control and experimental operation. GO could be an ideal substrate for growing and anchoring of functional nanocarystals for high-performance photocatalysts. Although decoration of titanium dioxide nanoparticles on GO sheets has been studied, [118, 119] it remains unexplored to control the properties of the nanocrystals growth on RGO by tuning the oxidation degree of the GO. Herein, we rationalize the nanocrystals growth behavior to enhance the photocatalytic performance of titanium dioxide/RGO composites.

References

1. R. Wang, X. Cai, F. Shen, Titanium dioxide hollow microspheres with mesoporous surface: superior adsorption performance for dye removal. Appl. Surf. Sci. **305**, 352–358 (2014)
2. J. Yu, L. Shi, One-pot hydrothermal synthesis and enhanced photocatalytic activity of trifluoroacetic acid modified titanium dioxide hollow microspheres. J. Mol. Catal. A: Chem. **326**, 8–14 (2010)
3. K. Qi, B. Cheng, J. Yu et al., Review on the improvement of the photocatalytic and antibacterial activities of ZnO. J. Alloy. Compd. **727**, 792–820 (2017)
4. S. Kuriakose, B. Satpati, S. Mohapatra, Enhanced photocatalytic activity of Co-doped ZnO nanodisks and nanorods prepared by a facile wet chemicalmethod. Phys. Chem. Chem. Phys. **16**, 12741–12749 (2014)

References

5. A. Hui, J. Ma, J. Liu, Y. Bao, J. Zhang, Morphological evolution of Fe doped search in-shaped ZnO nanoparticles with enhanced photocatalytic activity. J. Alloys Compd. **696**, 639–647 (2017)
6. L.C.-K. Liau, J.-S. Huang, Energy-level variations of Cu-doped ZnO fabricated through sol-gel processing. J. Alloys Compd. **702**, 153–160 (2017)
7. H. Benhebal, M. Chaib, A. Leonard, S.D. Lambert, M. Crine, Photodegradation of phenol and benzoic acid by solegel-synthesized alkali metal-doped ZnO. Mater. Sci. Semicond. Proc. **15**, 264–269 (2012)
8. B. Subash, B. Krishnakumar, R. Velmurugan, M. Swaminathan, M. Shanthi, Synthesis of Ce co-doped AgeZnO photocatalyst with excellent performance for NBB dye degradation under natural sunlight illumination. Catal. Sci. Technol. **2**, 2319–2326 (2012)
9. J.-C. Sin, S.-M. Lam, K.-T. Lee, A.R. Mohamed, Preparation of rare earth-doped ZnO hierarchical micro/nanospheres and their enhanced photocatalytic activity under visible light irradiation. Ceram. Int. **40**, 5431–5440 (2014)
10. S. Sharma, S.K. Mehta, S.K. Kansal, N doped ZnO/C-dots nanoflowers as visible light driven photocatalyst for the degradation of malachite green dye in aqueous phase. J. Alloys Compd. **699**, 323–333 (2017)
11. J.B.M. Goodall, D. Illsley, R. Lines, N.M. Makwana, J.A. Darr, Structure-property-Composition relationships in doped zinc oxides: enhanced photocatalytic activity with rare earth dopants. ACS Comb. Sci. **17**, 100–112 (2015)
12. D. Schelonka, J. Tolasz, V. Stengl, Doping of zinc oxide with selected first row transition metals for photocatalytic applications. Photochem. Photobiol. **91**, 1071–1077 (2015)
13. Q. Zhang, J.-K. Liu, J.-D. Wang, H.-X. Luo, Y. Lu, X.-H. Yang, Atmospheric selfinduction synthesis and enhanced visible light photocatalytic performance of Fe3þ doped Ag-ZnO mesocrystals. Ind. Eng. Chem. Res. **53**, 13236–13246 (2014)
14. R. He, R.K. Hocking, T. Tsuzuki, Co-doped ZnO nanopowders: location of cobalt and reduction in photocatalytic activity. Mater. Chem. Phys. **132**, 1035–1040 (2012)
15. Q. Yin, R. Qiao, Z. Li, X.L. Zhang, L. Zhu, Hierarchical nanostructures of nickel-doped zinc oxide: morphology controlled synthesis and enhanced visible light photocatalytic activity. J. Alloys Compd. **618**, 318–325 (2015)
16. R. Ullah, J. Dutta, Photocatalytic degradation of organic dyes with manganese-doped ZnO nanoparticles. J. Hazard. Mater. **156**, 194–200 (2008)
17. C.-J. Chang, T.-L. Yang, Y.-C. Weng, Synthesis and characterization of Cr-doped ZnO nanorod-array photocatalysts with improved activity. J. Solid State Chem. **214**, 101–107 (2014)
18. R. Slama, F. Ghribi, A. Houas, C. Barthou, L. El Mir, Visible photocatalytic properties of vanadium doped zinc oxide aerogel nanopowder. Thin Solid Films **519**, 5792–5795 (2011)
19. M. Fu, Y. Li, S. wu, P. Lu, J. Liu, F. Dong, Solegel preparation and enhanced photocatalytic performance of Cu-doped ZnO nanoparticles. Appl. Surf. Sci. **258**, 1587–1591 (2011)
20. N. Clament Sagaya Selvam, J.J. Vijaya, L.J. Kennedy, Effects of morphology and Zr doping on structural, optical, and photocatalytic properties of ZnO nanostructures. Ind. Eng. Chem. Res. **51**, 16333–16345 (2012)
21. D. Zhang, F. Zeng, Visible light-activated cadmium-doped ZnO nanostructured photocatalyst for the treatment of methylene blue dye. J. Mater. Sci. **47**, 2155–2161 (2011)
22. K. Kumar, M. Chitkara, I.S. Sandhu, D. Mehta, S. Kumar, Photocatalytic, optical and magnetic properties of Fe-doped ZnO nanoparticles prepared by chemical route. J. Alloys Compd. **588**, 681–689 (2014)
23. K. Selvam, M. Muruganandham, I. Muthuvel, M. Swaminathan, The influence of inorganic oxidants and metal ions on semiconductor sensitized photodegradation of 4-fluorophenol. Chem. Eng. J. **128**, 51–57 (2007)
24. L. Xu, Zhou Yang, Zijiun Wu, Zheng Gaige, J. He, Y. Zhou, Improved photocatalytic activity of nanocrystalline ZnO by coupling with CuO. J. Phys. Chem. Solids **106**, 29–36 (2017)
25. J. Iqbal, X. Liu, H. Zhu, Z.B. Wu, Y. Zhang, D. Yu, R. Yu, Raman and highlyultraviolet red-shifted near band-edge properties of LaCe-co-doped ZnO nanoparticles. Acta Mater. **57**, 4790–4796 (2009)

26. S. Anandan, A. Vinu, T. Mori, N. Gokulakrishnan, P. Srinivasu, V. Murugesan, K. Ariga, Photocatalytic degradation of 2, 4, 6-trichlorophenol using lanthanum doped ZnO in aqueous suspension. Catal. Commun. **8**, 1377–1382 (2007)
27. W. Zheng, Q. Miao, Y. Tang, W. Wei, J. Xu, X. Liu, Q. Qian, L. Xiao, B. Huang, Q. Chen, La(III)-doped ZnO/C nanofibers with coreeshell structure byelec-trospinning-calcination technology. Mater. Lett. **98**, 94–97 (2013)
28. M. Khatamian, A.A. Khandar, B. Divband, M. Haghighi, S. Ebrahimiasl, Heterogeneous photocatalytic degradation of 4-nitrophenol in aqueous suspension by Ln (La^{3+}, Nd^{3+} or Sm^{3+}) doped ZnO nanoparticles. J. Mol. Catal. AChem. **365**, 120–127 (2012)
29. V.H.-T. Thi, B.-K. Lee, Effective photocatalytic degradation of paracetamol using La-doped ZnO photocatalyst under visible light irradiation. Mater. Res. Bullet. **96**, 171–182 (2017)
30. K.-J. Kim, P.B. Kreider, C. Choi, C.-H. Chang, H.-G. Ahn, Visible-light-sensitive Na-doped p-type flower-like ZnO photocatalysts synthesized via a continuous flow microreactor. RSC Adv. **3**, 12702–12710 (2013)
31. A. Tabib, W. Bouslama, B. Sieber, A. Addad, H. Elhouichet, M. Ferid, R. Boukherroub, Structural and optical properties of Na doped ZnO nanocrystals: application to solar photocatalysis. Appl. Surf. Sci. **396**, 1528–1538 (2017)
32. D. Li, J.-F. Huang, L.-Y. Cao, H.-B. OuYang, J.-Y. Li, C.-Y. Yao, Microwave hydrothermal synthesis of Kþ doped ZnO nanoparticles with enhanced photocatalytic properties under visible-light. Mater. Lett. **118**, 17–20 (2014)
33. M. Yousefi, R. Azimirad, M. Amiri, A.Z. Moshfegh, Effect of annealing temperature on growth of Ce-ZnO nanocomposite thin films: X-ray photoelectron spectroscopy study. Thin Solid Films **520**, 721–725 (2011)
34. M. Rezaei, A. Habibi-Yangjeh, Microwave-assisted preparation of Ce-doped ZnO nanostructures as an efficient photocatalyst. Mater. Lett. **110**, 53–56 (2013)
35. V. Etacheri, R. Roshan, V. Kumar, Mg-doped ZnO nanoparticles for efficient sunlight-driven photocatalysis. ACS Appl. Mater. Interf. **4**, 2717–2725 (2012)
36. X. Qiu, L. Li, J. Zheng, J. Liu, X. Sun, G. Li, Origin of the enhanced photocatalytic activities of semiconductors: a case study of ZnO doped with Mg2þ. J. Phys. Chem. C **112**, 12242–12248 (2008)
37. A. Khataee, R. Darvishi Cheshmeh Soltani, Y. Hanifehpour, M. Safarpour, H. Gholipour Ranjbar, S.W. Joo, Synthesis and characterization of dysprosium-doped ZnO nanoparticles for photocatalysis of a textile dye under visible light irradiation. Ind. Eng. Chem. Res. **53**, 1924–1932 (2014)
38. P.V. Korake, A.N. Kadam, K.M. Garadkar, Photocatalytic activity of Eu^{3+}-doped ZnO nanorods synthesized via microwave assisted technique. J. Rare Earths **32**, 306–313 (2014)
39. M. Samadi, M. Zirak, A. Naseri, E. Khorashadizade, A.Z. Moshfegh, Recent progress on doped ZnO nanostructures for visible-light photocatalysis. Thin Solid Films **605**, 2–19 (2016)
40. S.R. Kadam, V.R. Mate, R.P. Panmand, L.K. Nikam, M.V. Kulkarni, R.S. Sonawane, B.B. Kale, A green process for efficient lignin (biomass) degradation and hydrogen production via water splitting using nanostructured C, N, S-doped ZnO under solar light. RSC Adv. **4**, 60626–60635 (2014)
41. L.-C. Chen, Y.-J. Tu, Y.-S. Wang, R.-S. Kan, C.-M. Huang, Characterization and photoreactivity of N-, S-, and C-doped ZnO under UV and visible light illumination. J. Photochem. Photobiol. A **199**, 170–178 (2008)
42. W. Yu, J. Zhang, T. Peng, New insight into the enhanced photocatalytic activity of N-, C- and S-doped ZnO photocatalysts. App. Cat. B En. **181**, 220–227 (2016)
43. X. Chen, Titanium dioxide nanomaterials and their energy applications. Chin. J. Catal. **30**, 839–851 (2009)
44. X. Chen, S.S. Mao, Titanium dioxide nanomaterials: synthesis, properties, modifications, and application. Chem. Rev. **107**(7), 2891–2959 (2007)
45. H. Zhang, Y. Song, Y. Sheng, H. Li, Z. Shi, X. Xu, H. Zou, EDTA-assisted fabrication of titanium dioxide core-shell microspheres with improved photocatalytic performance. Ceram. Int. **41**, 247–252 (2015)

References

46. N. Alenzi, W.-S. Liao, P.S. Cremer, V. Sanchez-Torres, T.K. Wood, C. Ehlig-Economides, Z. Cheng, Photoelectrochemical hydrogen production from water/methanol decomposition using Ag/titanium dioxide nanocomposite thin films. Int. J. Hydrog. Energy **35**, 11768–11775 (2010)
47. S.-S. Chen, H.-C. Hsi, S.-H. Nian, C.-H. Chiu, Synthesis of N-doped titanium dioxide photocatalyst for low-concentration elemental mercury removal under various gas conditions. Appl. Catal. B Environ. **160–161**, 558–565 (2014)
48. H. Li, C.-Y. Wu, Y. Li, J. Zhang, Superior activity of MnOx -CeO2/titanium dioxide catalyst for catalytic oxidation of elemental mercury at low flue gas temperatures. Appl. Catal. B: Environ. **111-112**, 381–388 (2012)
49. J. Yang, Q. Yang, J. Sun, Q. Liu, D. Zhao, W. Gao, L. Liu, Effects of mercury oxidation on V_2O_5 -WO_3/titanium dioxide catalyst properties in NH_3 -SCR process. Catal. Commun. **59**, 78–82 (2015)
50. H. Li, C.-Y. Wu, Y. Li, L. Li, Y. Zhao, J. Zhang, Impact of SO_2 on elemental mercury oxidation over CeO_2 titanium dioxide catalyst. Chcm. Eng. J. **219**, 319–326 (2013)
51. S. Xu, A.J. Du, J. Liu, J. Ng, D.D. Sun, Highly efficient CuO incorporated titanium dioxide nanotube photocatalyst for hydrogen production from water. Int. J. Hydrog. Energy **36**, 6560–6568 (2011)
52. L. Pan, S. Wang, J. Xie, L. Wang, X. Zhang, J.J. Zou, Constructing titanium dioxide p-n homojunction for photoelectrochemical and photocatalytic hydrogen generation. Nano Energy **28**, 296–303 (2016)
53. D.O. Scanlon, C.W. Dunnill, J. Buckeridge, S.A. Shevlin, A.J. Logsdail, S.M. Woodley, C.R. Catlow, M.J. Powell, R.G. Palgrave, I.P. Parkin, Band alignment of rutile and anatase titanium dioxide. Nature Mater. **12**, 798–801 (2013)
54. W. Zhou, W. Li, J.Q. Wang, Y. Qu, Y. Yang, Y. Xie, K. Zhang, L. Wang, H. Fu, D. Zhao, Ordered mesoporous black titanium dioxide as highly efficient hydrogen evolution photocatalyst. J. Am. Chem. Soc. **136**, 9280–9283 (2014)
55. C.H. Lee, J.L. Shie, Y.T. Yang, C.Y. Chang, Photoelectrochemical characteristics, photodegradation and kinetics of metal and non-metal elements co-doped photocatalyst for pollution removal. Chem. Eng. J. **303**, 477–488 (2016)
56. J. Zhang, X. Jin, P.I. Moralesguzman, X. Yu, H. Liu, H. Zhang, L. Razzari, J.P. Claverie, Engineering the absorption and field enhancement properties of Au-titanium dioxide nanohybrids via whispering gallery mode resonances for photocatalytic water splitting. ACS Nano **10**, 4496–4503 (2016)
57. M. Dahl, Y. Liu, Y. Yin, Composite titanium dioxide nanomaterials. Chem. Rev. **114**, 9853–9889 (2014)
58. J. Sun, X. Li, Q. Zhao, J. Ke, D. Zhang, Novel V_2O_5/$BiVO_4$/titanium dioxide nanocomposites with high visible-light-induced photocatalytic activity for the degradation of toluene. J. Phys. Chem. C **118**, 10113–10121 (2014)
59. S. Shen, S.A. Lindley, X. Chen, J.Z. Zhang, Hematite heterostructures for photoelectrochemical water splitting: rational materials design and charge carrier dynamics. Energy Environ. Sc. **9**, 2744–2755 (2016)
60. H. Huang, K. Xiao, Y. He, T. Zhang, F. Dong, X. Du, Y. Zhang, In situ assembly of BiOI@Bi12O17Cl2 p-n junction: charge induced unique front-lateral surfaces coupling heterostructure with high exposure of BiOI 001 active facets for robust and nonselective photocatalysis. Appl. Catal. B Environ. **199**, 75–86 (2016)
61. H. Huang, X. Han, X. Li, S. Wang, P.K. Chu, Y. Zhang, Fabrication of multiple heterojunctions with tunable visible-light-active photocatalytic reactivity in BiOBr-BiOI full-range composites based on microstructure modulation and band structures. ACS Appl. Mater. Interfaces. **7**, 482–492 (2015)
62. L. Xie, L. Ping, Z. Zheng, S. Weng, J. Huang, Morphology engineering of V_2O_5/titanium dioxide nanocomposites with enhanced visible light-driven photofunctions for arsenic removal. Appl. Catal. B Environ. **184**, 347–354 (2016)

63. F. Ren, H. Li, Y. Wang, J. Yang, Enhanced photocatalytic oxidation of propylene over V-doped titanium dioxide photocatalyst: Reaction mechanism between V^{5+} and single-electron-trapped oxygen vacancy. Appl. Catal. B Environ. **176–177**, 160–172 (2015)
64. Z. Wu, F. Dong, Y. Liu, H. Wang, Enhancement of the visible light photocatalytic performance of C-doped titanium dioxide by loading with V_2O_5. Catal. Commun. **11**, 82–86 (2009)
65. Y. Wang, Y.R. Su, L. Qiao, L.X. Liu, Q. Su, C.Q. Zhu, X.Q. Liu, Synthesis of one-dimensional titanium dioxide/V_2O_5 branched heterostructures and their visible light photocatalytic activity towards Rhodamine B. Nanotechnology **22**, 225702–225710 (2011)
66. Y. Duan, M. Zhang, L. Wang, F. Wang, L. Yang, X. Li, C. Wang, Plasmonic Ag-titanium dioxide$_{-x}$ nanocomposites for the photocatalytic removal of NO under visible light with high selectivity: the role of oxygen vacancies. Appl. Catal. B Environ. **204**, 67–77 (2016)
67. V. Gombac, L. Sordelli, T. Montini, J.J. Delgado, A. Adamski, G. Adami, M. Cargnello, S. Bernal, P. Fornasiero, CuO_x-titanium dioxide photocatalysts for H_2 production from ethanol and glycerol solutions. J. Phys. Chem. A **114**, 3916–3925 (2009)
68. W. Xu, H. Wang, X. Zhou, T. Zhu, CuO/titanium dioxide catalysts for gas-phase Hg^0 catalytic oxidation. Chem. Eng. J. **243**, 380–385 (2014)
69. V. Trevisan, A. Olivo, F. Pinna, M. Signoretto, F. Vindigni, G. Cerrato, C.L. Bianchi, C-N/titanium dioxide photocatalysts: Effect of co-doping on the catalytic performance under visible light. Appl. Catal. B Environ. **160–161**, 152–160 (2014)
70. L. Zhang, M.S. Tse, O.K. Tan, Y.X. Wang, M. Han, Facile fabrication and characterization of multi-type carbon-doped titanium dioxide for visible light-activated photocatalytic mineralization of gaseous toluene. J. Mater. Chem. A **1**, 4497–4507 (2013)
71. W. Zhao, Y. Wang, Y. Yang, J. Tang, Y. Yang, Carbon spheres supported visible-light-driven CuO-BiVO4 heterojunction: preparation, characterization, and photocatalytic properties. Appl. Catal. B: Environ. **115–116**, 90–99 (2012)
72. S. Xu, A.J. Du, J. Liu, J. Ng, D.D. Sun, Highly efficient CuO incorporated titanium dioxide nanotube photocatalyst for hydrogen production from water. Int. J. Hydrogen Energy **36**, 6560–6568 (2011)
73. J. Yu, Y. Hai, M. Jaroniec, Photocatalytic hydrogen production over CuO-modified titania. J. Colloid Interface Sci. **357**, 223–228 (2011)
74. J. Bandara, C.P.K. Udawatta, C.S.K. Rajapakse, Highly stable CuO incorporated titanium dioxide catalyst for photo-catalytic hydrogen production from H_2O. Photochem. Photobiol. Sci. **4**, 857–861 (2005)
75. D. Barreca, P. Fornasiero, A. Gasparotto, V. Gombac, C. Maccato, T. Montini, E. Tondello, The potential of supported Cu_2O and CuO nanosystems in photocatalytic H_2 production. Chemsuschem **2**, 230–233 (2009)
76. M. Khraisheh, L. Wu, A.A.H. Al-Muhtaseb, M.A. Al-Ghouti, Photocatalytic disinfection of *Escherichia coli* using titanium dioxide, P25 and Cu-doped titanium dioxide. J. Ind. Eng. Chem. **28**, 369–376 (2015)
77. H. Zangeneh, A.A.L. Zinatizadeh, M. Habibi, M. Akia, M.H. Isa, Photocatalytic oxidation of organic dyes and pollutants in wastewater using different modified titanium dioxides: a comparative review. J. Ind. Eng. Chem. **26**, 1–36 (2015)
78. A. Yousef, M.M. El-Halwany, N.A.M. Barakat, M.N. Al-Maghrabi, H.Y. Kim, CuO-doped titanium dioxide nanofibers as potential photocatalyst and antimicrobial agent. J. Ind. Eng. Chem. **26**, 251–258 (2015)
79. J. Jia, D. Li, J. Wan, X. Yu, Characterization and mechanism analysis of graphite/C-doped titanium dioxide composite for enhanced photocatalytic performance. J. Ind. Eng. Chem. **33**, 162–169 (2015)
80. M. Kong, Y. Li, X. Chen, T. Tian, P. Fang, F. Zheng, X. Zhao, Tuning the relative concentration ratio of bulk defects to surface defects in titanium dioxide nanocrystals leads to high photocatalytic efficiency. J. Am. Chem. Soc. **133**, 16414–16417 (2011)
81. Y. Xu, C. Zhang, L. Zhang, X. Zhang, H. Yao, J. Shi, Pd-catalyzed instant hydrogenation of titanium dioxide with enhanced photocatalytic performance. Energy Environ. Sci. **9**, 2410–2417 (2016)

82. H. Park, H. Kim, G. Moon, W. Choi, Photoinduced charge transfer processes in solar photocatalysis based on modified titanium dioxide. Energy Environ. Sci. **9**, 411–433 (2015)
83. C. Li, C. Koenigsmann, W. Ding, B. Rudshteyn, K.R. Yang, K.P. Regan, S.J. Konezny, V.S. Batista, G.W. Brudvig, C.A. Schmuttenmaer, Facet-dependent photoelectrochemical performance of titanium dioxide nanostructures: an experimental and computational study. J. Am. Chem. Soc. **137**, 1520–1529 (2015)
84. J. Yun, S. Hwang, J. Jang, Fabrication of Au@Ag core/shell nanoparticles decorated titanium dioxide hollow structure for efficient light-harvesting in dye-sensitized solar cells. ACS Appl. Mater. Interfaces. **7**, 2055–2063 (2015)
85. W.J. Lee, M.L. Ju, S.T. Kochuveedu, T.H. Han, Y.J. Hu, M. Park, J.M. Yun, J. Kwon, K. No, H.K. Dong, Biomineralized N-doped CNT/titanium dioxide core/shell nanowires for visible light photocatalysis. ACS Nano **6**, 935–943 (2012)
86. R. Chalasani, S. Vasudevan, Cyclodextrin-functionalized Fe_3O_4@titanium dioxide: reusable, magnetic nanoparticles for photocatalytic degradation of endocrine-disrupting chemicals in water supplies. ACS Nano **7**, 4093–4104 (2013)
87. Y.C. Chen, Y.C. Pu, Y.J. Hsu, Interfacial charge carrier dynamics of the three-component In_2O_3-titanium dioxide-Pt heterojunction system. J. Phys. Chem. C **116**, 2967–2975 (2012)
88. F. Lei, Y. Sun, K. Liu, S. Gao, L. Liang, B. Pan, Y. Xie, Oxygen vacancies confined in ultrathin indium oxide porous sheets for promoted visible-light water splitting. J. Am. Chem. Soc. **136**, 6826–6829 (2014)
89. J. Mu, B. Chen, M. Zhang, Z. Guo, Z. Peng, Z. Zhang, Y. Sun, C. Shao, Y. Liu, Enhancement of the visible-light photocatalytic activity of In_2O_3-titanium dioxide nanofiber heteroarchitectures. ACS Appl. Mater. Interfaces. **4**, 424–430 (2016)
90. R. Asahi, T. Morikawa, H. Irie, T. Ohwaki, Nitrogen-doped titanium dioxide as visible-light-sensitive photocatalyst: designs, developments, and prospects. Chem. Rev. **114**, 9824–9852 (2014)
91. J. Zhang, N.M. Vasei, Y. Sang, H. Liu, J.P. Claverie, Titanium dioxide@Carbon photocatalysts: the effect of carbon thickness on catalysis. ACS Appl. Mater. Interfaces. **2**, 1903–1912 (2015)
92. S.S. Dr, H.K. Dr, Daylight photocatalysis by carbon-modified titanium dioxide. Angew. Chem. Int. Ed. **42**, 4908–4911 (2003)
93. W. Jiang, C.E. Li, X. Chen, J. Zhang, L. Zhao, T. Huang, T. Hu, C. Zhang, B. Ni, X. Zhou, Photocatalytic oxidation of gas-phase Hg^0 by carbon spheres supported visible-light-driven CuO-titanium dioxide. J. Indust. Eng. Chem. **46**, 416–425 (2016)
94. S. Chala, K. Wetchakun, S. Phanichphant, B. Inceesungvorn, N. Wetchakun, Enhanced visible-light-response photocatalytic degradation of methylene blue on Fe-loaded $BiVO_4$ photocatalyst. J. Alloy. Compd. **597**, 129–135 (2014)
95. X. Gao, F. Fu, W. Li, Photocatalytic degradation of phenol over Cu loading $BiVO_4$ metal composite oxides under visible light irradiation. Physica B: Condensed Matter. **412**, 26–31 (2013)
96. H. Xu, C. Wu, H. Li, J. Chu, G. Sun, Y. Xu, Synthesis, characterization and photocatalytic activities of rare earth-loaded $BiVO_4$ catalysts. Appl. Surf. Sci. **256**(3), 597–602 (2009)
97. A. Zhang, J. Zhang, Effects of europium doping on the photocatalytic behavior of $BiVO_4$. J. Hazard. Mater. **173**(1/2/3), 265–272 (2010)
98. J. Li, Z. Guo, H. Liu, J. Du, Z. Zhu, Two-step hydrothermal process for synthesis of F-doped $BiVO_4$ spheres with enhanced photocatalytic activity. J. Alloy. Compd. **581**, 40–45 (2013)
99. R. Leary, A. Westwood, Carbonaceous nanomaterials for the enhancement of titanium dioxide photocatalysis. Carbon **49**, 741–772 (2011)
100. A. Fujishima, K. Hashimoto, T. Watanabe, Titanium dioxide photocatalysis and related surface phenomena. Surf. Sci. Rep. **63**, 515–582 (2008)
101. K. Hashimoto, H. Irie, A. Fujishima, Photocatalysis: a historical overview and future prospects. Jpn. J. Appl. Phys. **44**, 8269 (2005)
102. X. Chen, S.S. Mao, Titanium dioxide nanomaterials synthesis, properties, modifications, and applications. Chem. Rev. **107**, 2891–2959 (2007)

103. A. Heller, Chemistry and applications of photocatalytic oxidation of thin organic films. Acc. Chem. Res. **28**, 503–508 (1995)
104. Z. Zhang, C. Shao, L. Zhang, X. Li, Y. Liu, Electrospun nanofibers of V-doped titanium dioxide with high photocatalytic activity. J. Colloid Interface Sci. **351**, 57–62 (2010)
105. M.-Z. Ge, S.-H. Li, J.-Y. Huang, K.-Q. Zhang, S.S. Al-Deyab, Y.-K. Lai, Titanium dioxide nanotube arrays loaded with reduced graphene oxide films: facile hybridization and promising photocatalytic application. J. Mater. Chem. A **3**, 3491–3499 (2015)
106. Q. Zhang, J.-B. Joo, Z. Lu, M. Dahl, D.Q. Oliveira, M. Ye, Y. Yin, Self-assembly and photocatalysis of mesoporous titanium dioxide nanocrystal clusters. Nano Res. **4**, 103–114 (2011)
107. B.J. Ji, Q. Zhang, M. Dahl, I. Lee, J. Goebl, F. Zaera, Y. Yin, Control of the nanoscale crystallinity in mesoporous titanium dioxide shells for enhanced photocatalytic activity. Energy Environ. Sci. **5**, 6321–6327 (2012)
108. I. Robel, B.A. Bunker, P.V. Kamat, Single-walled carbon nanotube-CdS nanocomposites as light-harvesting assemblies: photoinduced charge-transfer interactions. Adv. Mater. **17**, 2458–2463 (2005)
109. A.K. Geim, K.S. Novoselov, NS the rise of graphene. Nat. Mater. **6**, 183–191 (2007)
110. H. Wang, J.T. Robinson, G. Diankov, H. Dai, Nanocrystal growth on graphene with various degrees of oxidation. J. Am. Chem. Soc. **132**, 3270–3271 (2010)
111. J. Zhang, Z. Zhu, Y. Tang, X. Feng, Graphene encapsulated hollow titanium dioxide nanospheres: efficient synthesis and enhanced photocatalytic activity. J. Mater. Chem. A **1**, 3752–3756 (2013)
112. K. Woan, G. Pyrgiotakis, W. Sigmund, Photocatalytic carbon-nanotube-titanium dioxide Composites. Adv. Mater. **21**, 2233–2239 (2009)
113. H. Zhang, X. Lv, Y. Li, Y. Wang, J. Li, P25-graphene composite as a high performance photocatalyst. ACS Nano **4**, 380–386 (2009)
114. C.H. Kim, B.-H. Kim, K.S. Yang, Titanium dioxide nanoparticles loaded on graphene/carbon composite nanofibers by electrospinning for increased photocatalysis. Carbon **50**, 2472–2481 (2012)
115. B.C. Brodie, On the Atomic Weight of Graphite. Philos. Trans. R. Soc. Lond. 249–259 (1859)
116. L. Staudenmaier, Verfahren zur Darstellung der Graphitsäure. Ber. Dtsch. Chem. Ges. **31**, 1481–1487 (1898)
117. W.S. Hummers Jr., R.E. Offeman, Functionalized graphene and graphene oxide: materials synthesis and electronic applications. J. Am. Chem. Soc. **80**, 1339 (1958)
118. C. Chen, W. Cai, M. Long, B. Zhou, Y. Wu, D. Wu, Y. Feng, Synthesis of visible-light responsive graphene oxide/titanium dioxide composites with p/n heterojunction. ACS Nano **4**, 6425–6432 (2010)
119. G. Jiang, Z. Lin, C. Chen, L. Zhu, Q. Chang, N. Wang, W. Wei, H. Tang, Titanium dioxide nanoparticles assembled on graphene oxide nanosheets with high photocatalytic activity for removal of pollutants. Carbon **49**, 2693–2701 (2011)

Chapter 5
Photocatalytic Denitrification in Flue Gas

Abstract NO_X and SO_2 emission from power plants during the coal burning process has been one of the major problems that result in adverse effect on the environment and human health. In general, ammonia based selective catalytic reduction (NH_3-SCR) and calcium-based wet flue gas desulfurization (WFGD-Ca) processes have been applied to flue gas treatment of coal-fired power plants on a large scale, but they have not been able to achieve the comprehensive removal of a variety of pollutants. The combination of NH_3-SCR and WFGD-Ca processes is mainly used to remove both simultaneous NO_X and SO_2, but the high capital operating costs limit its use in developing countries. In recent years, the study on the degradation of wastewater and gaseous pollutants by titanium dioxide has received extensive attention and has obtained good results in basic research and application-oriental research. Fe_3O_4-titanium dioxide composites were prepared by hydrothermal method, and the titanium dioxide layer was coated on the surface of Fe_3O_4. Because of the benefits of photocatalytic method, it has been paid attention to control NOx and other pollutant in the flue gas. The main principle of photocatalytically treating NO_x is to photocatalytically oxidize NO into NO_2, which is soluble and may be captured by WFGD through liquid absorption.

Keywords NO_X · Photocatalytic · Fe_3O_4-titanium dioxide · NO_x removal

5.1 Denitrification in the Flue Gas

5.1.1 The Importance of Denitration

With the fast development of the economy in this world, the energy and coal consumption becomes more and more, especially in China the coal-based energy accounts for nearly 70% of total energy consumption. Due to the lack of oil and gas, coal is its basic electricity generation resource. Therefore, China's power industry is based on coal-combustion power generation. On the basis of the statistics of nitrogen oxides (NO_x) emissions in China's power industry in recent years ('National Environmental Bulletin (2014)'), there were 154,633 industrial enterprises in China, and the nitrogen oxides (NO_x)emission was 14.048 million tons. Among them, the number of thermal power industries was 3288, and a total amount of 8.83 million tons of nitrogen oxides (NO_x) were emitted, accounting for 55.62% of the total emissions. According to statistic data, nitrogen oxides (NO_x) emissions increased year by year, coal-combustion power plant pollution problems became more and more serious, and if it is not effectually controlled, it will bring thoughtful environmental damage.

Nitrogen oxides (NO_x) is a highly toxic pollutant, which can destroy the ozone layer, ecological environment and human health. NO_x is mostly produced by the coal, oil, natural gas and other petrochemical fuel combustion. NO_x is composed of NO and NO_2, in which NO accounts for more than 90%. NO_x has strong affinity for hemoglobin, once it enters the blood and hemoglobin to resolutely combine with the blood hypoxia, not only causing bronchitis, emphysema and other diseases, but also improving premature aging, bronchial epithelial cell lymphoid tissue hyperplasia, and even lung cancer symptom. Nitrogen oxide is one of the major atmospheric pollutants. It is one of the main reasons causing acid rain, ozone hole, photochemical smog and other environmental problems and ecological hazards. NO_x emissions to the atmosphere will result in the formation of O_3 and photochemical smog, leading to human eye swelling, sore throat cough, skin flushing or even heart and lung failure and other symptoms. In addition, it can have serious influence on human health, irritating the nose, throat and lungs, and can easily damage the human respiratory system, causing bronchitis and other diseases. NO_x and water will form HNO_3 and HNO_2, leading to formation of acid rain. NO_x is also harmful to plants, causing leaf necrosis, whitening, yellowing or brown spots.

Emissions from coal-fired flue gas of sulfur dioxide (SO_2) and oxides of nitrogen (NO_x) are not only significant precursors of haze in the main pollutant $PM_{2.5}$, but also lead to acid rain, smog, ozone layer destruction and other serious pollution problems, having a serious impact on the environment and human health [1, 2]. The air pollution prevention and control of environmental monitoring of pollutant emission in thermal power plants have been paid attention to, and the emission targets have been proposed and the related laws and regulations have been issued and NO_x, SO_2 emission index declines year by year. The NO_x, SO_2 emission standard of coal-fired power plant becomes more and more strict, and environmental protection equipment in power plants and emission reduction technology have been put forward higher request. The enterprise has to implement the strategy of sustainable development, and the coal-combustion power plant must adopt advanced environmental protection device and technology to reduce emissions. The coal-consumption power enterprises must strengthen energy conservation and emission reduction technology, effectively controlling pollutant discharge to improve environmental protection.

At the moment, countries all over the world have capitalized lots of manpower and material resources to develop a series of flue gas purification technology competing in desulfurization and denitrification aspects, such as wet, dry, semi-dry desulfurization, denitrification selective catalytic reduction (SCR), and activated carbon adsorption, electron beam radiation, corona, wet desulfurization denitration technology, and etc. [3]. Among them, the most mature and widely used technique for controlling SO_2

5.1 Denitrification in the Flue Gas

in flue gas is wet limestone/gypsum flue gas desulfurization (WFGD). For NO_x, at present, the control technology of NO_x pollution produced by combustion is divided into two major categories: pre-production control (fuel denitrification, low NO_x combustion process), and post-production flue gas denitrification. The fuel denitrification technology is not yet of a good development, while low NO_x combustion technology at abroad and home have done lots of study, and some technologies and equipments have been developed and adopted in industrial application, but the denitration efficiency is not enough to meet the strict NO_x emission regulation. At present, flue gas denitration is still the main technology of NO_x pollution control, and selective catalytic reduction process denitration (SCR) is mostly used [4]. Although there are still some defects in aspects of technology and economy in the dry flue gas instantaneous desulfurization denitration process, it remains to be a researched flue gas purification technology [5] due to the advantages of low water consumption, low operation cost, simple equipment, lower land consumption, ease of sulfur recovery, etc. Photocatalytical NO_x removal technology and relative technologies are one of important NO_x removal methods because of its being without secondary pollutant, and it has brought more and more attention.

5.1.2 Photocatalytic Technology

Semiconductor photocatalysis is the most active field in many researches on pollution control by photochemical method, which has become a hotspot in the research of pollution control chemistry and has formed a new research field. Nowadays, the study on the degradation of wastewater and gaseous pollutants by titanium dioxide has gotten widely attention and has attained good results in fundamental studies and application-oriental research. Titanium dioxide is very stable because of its chemical and photochemical properties, and it is non-toxic and inexpensive. Therefore, in photocatalytic removal, titanium dioxide is usually adopted as a photocatalyst. Photocatalyst redox mechanism is mainly affected by photocatalyst itself, absorbed-light energy, electron transition occurring, electronic-hole pair generation, and the pollutants adsorbed upon the surface of the photocatalyst. The straight oxidation of hydroxyl OH^- absorbed on the surface of the photocatalyst generates hydroxyl radical ·OH with strong oxidizing ability, which will in turn oxidize pollutants.

Titanium dioxide, with excellent catalytic performance, is not only a kind of new catalytic material that has wide application prospects in the field of environmental protection, but also one of the worldwide studied hotspots in the field of new catalytic materials. Nano titanium dioxide has the advantages of no toxicity, low cost, mild reaction conditions, stable chemical properties and etc. [6] With the small particle size, high surface activity, a unique surface effect and quantum effect, titanium dioxide has been broadly used in wastewater treatment, air purification and other fields [7]. However, there are still many factors limiting its industrial application [8]. For example, as a broad band gap semiconductor, it can only be stimulated in the ultraviolet light [9], and it is with a low electron transfer rate and high recombination rate of photogenerated electron-hole pairs [10]. In order to address these challenges, finding a novel photocatalyst which could be excited under visible light and could produce high quantum yield has been one of the research hotspots by the scientists. At the same time, it is also an important issue for NO_x removal in environment treatment with a suitable photocatalyst being environment-friendly and efficient.

5.1.2.1 Photocatalytic Oxidation/Reduction Mechanism

Photocatalysis is based on the band theory of n-type semiconductor, with n-type semiconductor as a sensitizer of a photosensitive oxidation method. The most commonly n-type semiconductors used as photocatalysts are metal oxides and sulfides, such as ZnO, CdS, Fe_2O_3, SnO_2 and WO_3. The semiconductor band is discrete structure and often consists of a low-power valence band being full of electrons (valence band, VB) and an empty high energy conduction band (conduction band, CB). An area between the valence band and conduction band is the band gap, and the size of the area is often called the forbidden band width (Eg) [11].

When the energy is equal to or bigger than the band width (band gap energy, Eg) of the semiconductor, the semiconductor absorbs photon energy, and the electron transition of valence band to the conduction band is called interband transition [12]. When an electron is excited from the valence band to the conduction band, a high active electron (e^-) with negative charge is generated on the conduction band, and a hole (h^+) with positive charge is left in the valence band, forming an electron-hole pair [13, 14]. Semiconductor particles are different from metal in that the former is lack of continuous areas between the energy band, and electronic-hole, so semiconductor particles have a long service life [15]. Under the effect of potential field, the electrons continuously enter the empty zone without electrons from one direction, which causes the hole to move the opposite direction [16]. That is to say, negatively charged electrons move to one direction, corresponding to positively charged holes moving to the opposite direction, so that they can migrate to the surface of semiconductor particles. The electrons and holes that reach the surface of the semiconductor particle can go through two separate processes. On the surface, the semiconductor easily transfers electrons to the electron acceptor and reduces the acceptor, while the holes can migrate to the surface and combine with the electron-donating species to oxidize the species. The process of competition with the charge carrier migration is the combination of electron and hole.

5.1.2.2 Photocatalytic Denitrification Classification

The photocatalytic denitrification may be classified into two kinds of methods. One is that photocatalysis is used to make NO_x oxidation. The hole of semiconductor material has strong electron-gaining ability. When the light stimulates electrons and holes, the electrons in the NO_x system can be captured, and NO_x is activated and oxidized. The oxygen in water and air and the electrons in the catalytic material react to produce the strong oxidation of · OH and · O_2^- free radicals. These free radicals can eventually oxidize NO_x to NO_3^-. NO_3^- will remain on the surface of photocatalyst, and when it is accumulated to a certain amount, the activity of the catalyst will be reduced and then become ineffective. Therefore, it is necessary to wash the surface of the catalyst regularly to regenerate it. After washing the catalyst with water, it can be neutralized completely without causing secondary pollution.

The other is photocatalytic reduction NO_x. Photocatalysis is also the principle of photocatalyst, and NO_x is directly decomposed into completely harmless nitrogen. Compared with photocatalytic oxidation, the photocatalytic reduction reaction has no accumulated NO_3^-, so it is no need to clean the surface, and no need to regenerate.

5.1.2.3 Factors that Affect Photocatalytic Activity

The reaction rate is related to the photocatalytic technology, it is an important factor which can be put into practical application of that in semiconductor photocatalytic reaction system. The photocatalytic activity is the main factor to determine the photocatalytic reaction rate speed, and it is mainly affected by catalyst crystal structure, catalyst dosing quantity, pollutant concentration, reaction temperature, pH of the solution, light source and light intensity, etc.

There are two main types of titanium dioxide for photocatalysis: anatase and rutile [17]. In the photocatalytic reaction, the photocatalytic activity of anatase TiO_2 is higher than that of rutile TiO_2. The particle size of the catalyst directly influences the photocatalytic activity. The smaller the particle size, the more the particle number per unit mass, the larger the surface area. For general multiphase catalytic reaction, when the density of the active center on the catalyst surface is constant, the catalyst with larger surface area is more active [18–20]. For photocatalytic reaction, it is caused by photogenic electrons and holes to adsorb OH and produce a highly reactive · OH, · OH can oxidize the reactants. The surface of the photocatalyst is an important factor to determine the adsorption amount of the reaction matrix since large surface area of catalyst means large adsorption, which is beneficial to photocatalytic reaction on the surface and shows high activity. At the same time, other factors, such as lattice defect and active centers, are also important for the photocatalytic reaction. Because of its energy band, the excitation light of titanium dioxide as photocatalyst is the ultraviolet light or near ultraviolet part. With modification or doped with metal, nonmetal, its excitation light source may extend to visible light with sunshine or artificial light source such as sterilization lamp, mercury lamp and xenon lamp, which light intensity range is large and the wavelength can be adjustable, so its application is very convenient.

5.1.2.4 Preparation of Titanium Dioxide Photocatalyst

In recent years, due to the quantum effects of nanomaterials, compared to block material, nano titanium dioxide has the advantages of large specific surface area, wide range of suction light, relatively low rate of electronic-hole recombination and corresponding high redox potential [21]. Therefore, besides the development of nano titanium dioxide with higher quantum yield, significant size effect and surface effect is also an important research direction for many scholars. The preparation methods of nano titanium dioxide includes hydrothermal method, hydrolysis method, microemulsion method, sol-gel method, homogeneous precipitation method, laser chemistry method, plasma method and bright ion beam method. Some scholars also adopt physical crushing method, sputtering method, electrolysis method, explosion method and spray pyrolysis method to prepare nano TiO_2. Among them, the hydrolysis method and sol-gel method [22] have been widely studied, which is a commonly method in current research.

The simplest way to prepare nano titanium dioxide is to direct hydrolysis of titanium salt solution. It is also one of the commonly used preparation methods. Titanium alkoxide, especially the four titanium isopropyl oxygen radicals are the most common materials in the preparation of titanium dioxide sol, often using ethanol, normal propyl alcohol as solvent, hydrolysis of nitric acid, acetic acid and other inorganic acid as the catalyst system. Hydrolysis is often carried out by adding acetone, methyl cellulose and other organic auxiliary systems to make the sol disperse more evenly. Phthalate esters can form amorphous TiO_2 by slow hydrolysis at room temperature or at a certain temperature, and then be transformed to catalytic active anatase titanium dioxide or titanium dioxide transparent film at high temperature.

The coating method based on sol-gel method is the most widely studied method for preparing titanium dioxide film. Compared with chemical vapor deposition and sputtering film, sol-gel method is with low operating temperature and is easy to be applied. The basic steps of this method is the preparation of sol-gel solution first, and then use dip coating, spin coating or spray sol solution to the substrate, and then base material dry roasting, getting attached to the titanium dioxide film on the surface of the base material.

5.2 Photocatalytic Denitrification in Flue Gas

In recent years, the research on the degradation of waste-water and gaseous pollutants by titanium dioxide (TiO_2) has received extensive attention and has achieved some results in basic and applied research. Titanium dioxide, with excellent catalytic performance, is not only a kind of new catalytic material that has wide application prospects in the field of environmental protection, but also one of the worldwide research hotspots in the field of new catalytic materials. Nano titanium dioxide has the advantages of no toxicity, low cost, mild reaction conditions, stable chemical properties and so on [6]; and with the small particle size, high surface activity, a unique surface effect and quantum effect, it has been widely used in wastewater treatment, air purification and other fields [7]. However, due to the aggregation of titanium dioxide nanoparticles in the reaction system, the effective specific surface area of the catalyst subsequently decreases, which leads to the rapid decrease of

5.2 Photocatalytic Denitrification in Flue Gas

the activity. Nevertheless, the use of such ultrafine particles will also cause a lot of separation issue [8]. In order to solve the issue of titanium dioxide powder, for example, easy to gather and difficult to recycle, to make full use of the characteristics of nano titanium dioxide and to enhance its activity, the titanium dioxide composite material was studied [9, 10, 23]. Scholars tried to fix the titanium dioxide powder on the magnetic carrier to prepare a magnetically recyclable titanium dioxide composite particle catalyst. The material can be recovered and reused under the external magnetic field [24, 25]. This magnetic catalyst has high catalytic performance and is easy for magnetic separation and recovery. Magnetic carrier Fe_3O_4 magnetic particle is commonly used worldwide. Chen [26] synthesized flower-like Fe_3O_4-titanium dioxide core shell nano particles. However, there are few research on simultaneous desulfurization and denitrification of Fe_3O_4-titanium dioxide composites, and the principle of flue gas simultaneous desulfurization and denitrification based on the Fe_3O_4-titanium dioxide is less reported. The hydrothermal method was applied to prepare Fe_3O_4-titanium dioxide composite, in which titanium dioxide layer was wrapped on the surface of Fe_3O_4. The reaction activity of different proportions of adsorbent under different reaction temperatures were studied with simulated flue gas flow rate and temperature to research desulfurization and denitration performance of the material.

5.2.1 Experimental

5.2.1.1 Experimental Reagents and Instruments

Laboratory reagent: tetrabutyl titanate (TBOT), anhydrous ethanol, triethylamine (TEA), iron oxide (Fe_3O_4) were of analytical grade, purchased from Sinopharm Chemical Reagent Co., Ltd. Instruments used in the experiments include: JEM-2010 type HRTEM analyzer, Japan Jeol company; Magna-750 Fourier transform infrared spectrometer, US Nicolet company; 101-3AB type blast electric dryer, Experimental instrument Co., Ltd. of North China in Tianjin; KM9106 flue gas analyzer, British Kane company.

5.2.1.2 Preparation for Fe_3O_4-Titanium Dioxide Composites

Fe_3O_4-titanium dioxide was synthesized by hydrothermal method with tetrabutyl titanate as titanium source. According to the mass ratio, the corresponding purchased Fe_3O_4 were put into the four beakers respectively, being added 30 mL of anhydrous ethanol, 10 mL tetrabutyl titanate, and suitable amount of deionized water, with ultrasonic dispersion for 15 min. The reaction solution was then transferred into a high pressure reaction kettle with a 250 mL lined PTFE, in which a small amount of deionized water is added. The reaction kettle was then placed in an oven at 150 °C for 4 h, cooled to room temperature naturally. Products were washed several times with anhydrous ethanol and deionized water, then placed in a vacuum drying oven at 110 °C for 12 h, set aside. The material was recorded as x% Fe_3O_4-titanium dioxide, where x% was the mass fraction of Fe_3O_4. At the same time, pure titanium dioxide nano materials were prepared by the same method.

5.2.1.3 Characterization of Materials

The simulated flue gas of $NO/SO_2/O_2/N_2$ in the experiment was generated by the control of the dynamic air distribution system. The inner diameter of the quartz tube fixed-bed reactor was 7 mm, the gas flow rate 200 mL/min, the loading capacity 0.35 g, and space velocity 24,000 h^{-1}. The import feed gas mixture contained 500 ppm SO_2, 300 ppm NO, 5% O_2, N_2 as the balance gas and the reaction temperature was controlled by the program temperature control device. When the outlet NO gas concentration reached 80% of imports on the adsorption bed, it was defined to reach the point of penetration. The concentrations of SO_2, NO_x (NO and NO_2) in flue gas were measured by KM9106 portable integrated flue gas analyzer (Kane). The removal rate and adsorption capacity of SO_2 and NO_x were calculated according to the concentration of SO_2, NO and NO_2 of the product. The removal rate was defined as:

$$\eta = \frac{C_{inlet} - C_{outlet}}{C_{inlet}} \times 100\% \tag{5.1}$$

where C_{inlet} was the concentration (ppm) at the inlet of the reactor, C_{outlet} was the concentration at the outlet.

The formula for calculating the adsorption capacity was:

$$S = \frac{\int_0^T (C_{inlet} - C_{outlet}) Q dt}{G} \tag{5.2}$$

where S was the saturated adsorption capacity (mmol g^{-1}), t was the adsorption time (min), C_{inlet} was the concentration (ppm) at the inlet of the reactor, C_{outlet} was the concentration (ppm) at the outlet, Q was the feed flow(mL min^{-1}), and G was the mass of adsorbent.

5.2.2 Results and Discussion

5.2.2.1 Comparison of the Performance of Fe_3O_4-Titanium Dioxide, Fe_3O_4 and Titanium Dioxide on Simultaneous Desulfurization and Denitrification

The samples of the prepared titanium dioxide, the purchased of raw materials Fe_3O_4, 10% Fe_3O_4-titanium dioxide were used as adsorbent at 100 °C in the simul-

5.2 Photocatalytic Denitrification in Flue Gas

taneous desulfurization and denitrification experiment. In order to compare the difference in performance between the composite material and the physically mixed components, the samples of the same quality of the prepared titanium dioxide and the purchased raw materials Fe_3O_4 mixed physically were tested under the similar conditions. Shown as Fig. 5.1, the activity of Fe_3O_4 was poor and SO_2 and NO_x were very quick to penetrate. The activity of titanium dioxide on simultaneous desulfurization and denitrification was better. There was no SO_2 at the outlet within the measured time, and NO_x removal rate was reduced to 20% after 56 min. In contrast, the activity of 10% Fe_3O_4-titanium dioxide was higher, in which SO_2 was not detected at the outlet until 64 min, while NO_x decreased to 20%, and the overall effect of 10% Fe_3O_4-titanium dioxide was higher than titanium dioxide. The desulfurization effect of Fe_3O_4-titanium dioxide, Fe_3O_4 and titanium dioxide mixed physically were basically the same, achieving almost 100% removal efficiency in the testing time. But the difference on denitrification was large, physical mixture of adsorbent was close to penetrate around 32 min, while the removal rate of Fe_3O_4-titanium dioxide composite material was reduced to 20% around 64 min. In order to compare the adsorption effect of different materials, the adsorption capacity of each material was calculated. When NO concentration at the outlet reached 80% of the concentration at the inlet of the adsorption bed, it was deemed to reach the breakthrough point corresponding to the absorption time. The adsorption capacity of each material was summarized in Table 5.1: Fe_3O_4-titanium dioxide material had the strongest adsorption capacity, and the adsorption capacity of NO and SO_2 were 0.2031 and 0.8152 mmol g^{-1} respectively.

With the same quantity, the activity of Fe_3O_4-titanium dioxide material was higher than that of Fe_3O_4 and pure titanium dioxide. It shows that Fe_3O_4-titanium dioxide is a kind of special composite structure, and Fe_3O_4-titanium dioxide has good dispersibility and stability with the capability of improving the material of simultaneous desulfurization and denitrification activity. By the comparison of Fe_3O_4-titanium

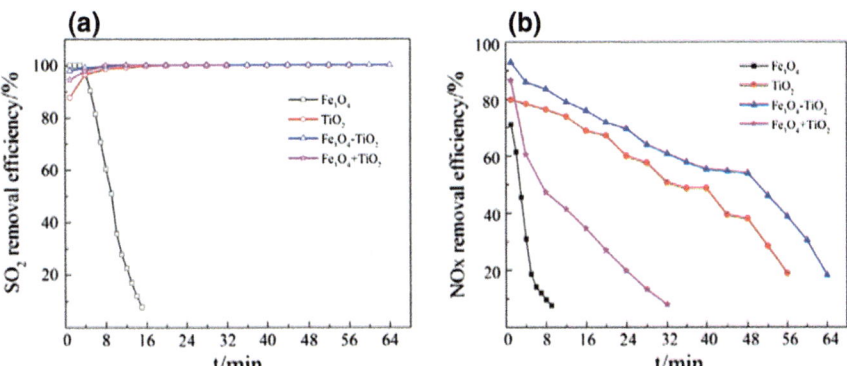

Fig. 5.1 Comparison of the performance of Fe_3O_4-titanium dioxide, Fe_3O_4 and titanium dioxide on simultaneous desulfurization and denitrification. Reprinted from Ref. [27], Copyright 2016, with permission from Elsevier

Table 5.1 Adsorption capacity of NO_x and SO_2 on different adsorbents

Adsorption capacities (mmol g^{-1})	Fe_3O_4	TiO_2	Fe_3O_4–TiO_2	Fe_3O_4 + TiO_2
S(NO)	0.014	0.1556	0.2031	0.0457
S(SO$_2$)	0.0629	0.7092	0.8152	0.3041

Reprinted from Ref. [27], Copyright 2015, with permission from Elsevier

dioxide with the physical mixture of Fe_3O_4 and titanium dioxide, it also can be seen that Fe_3O_4-titanium dioxide is not a simple physical mixture, because it may form a special package structure. The surface layer of nano titanium dioxide and Fe_3O_4 carrier promote each other, improving the material simultaneous desulfurization and denitrification activity. The Fe_3O_4 itself does not have a high catalytic activity, only acting as the role of a promoter. It can promote NO to be better affected with the material's surface to strengthen its chemical adsorption process [28], then cause more sulfur dioxide and oxynitride to be adsorbed on the surface of titanium dioxide. The addition of metal oxides can significantly improve the catalytic activity of the catalyst, and these metal oxides include Fe_3O_4, Al_2O_3, CaO, MgO, and etc. [29, 30].

5.2.2.2 Study on Simultaneous Desulfurization and Denitration of Fe_3O_4-Titanium Dioxide Composites at Different Reaction Temperatures

The 10% Fe_3O_4-titanium dioxide was adopted to conduct the simultaneous desulfurization and denitrification experiments at different reaction temperatures. In Fig. 5.2a, there was a slight gap in the first few minutes at each temperature in terms of desulfurization. With the increase of reaction time, the desulfurization efficiency of the two kinds of materials under various temperature both can reach almost 100% and the denitrification effect decreased particularly with the rise of temperature. This could be explained that the NO_x is reacting on the surface of the Fe_3O_4-titanium dioxide adsorbent, including both physical and chemical adsorption. The physical adsorption process is caused by the van der Waals forces between NO_x molecules and the surface of the titanium dioxide particles, while the physical process is reversible and can cause the phenomenon that the heated molecular escapes from the surface of the adsorbent. Due to the presence of sulfur dioxide in the experiment, the reactant particle diffusion rate and the collision frequency are improved with the increase of temperature, and the reaction rate is accelerated, which leads to increased chances of the molecule to spread on the surface of the adsorbent particles and to occupy more adsorption holes [31], so that on a certain amount of load catalyst, it reduces the adsorption of the Fe_3O_4-titanium dioxide particles on NO_x molecular.

From Fig. 5.2a, b, it can be seen that two kinds of materials are with higher removal efficiency of SO_2 than that of NO_x. This may be due to the relevant adsorption

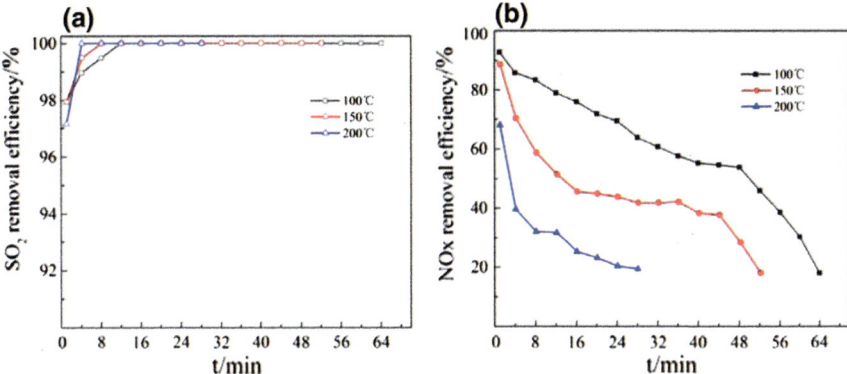

Fig. 5.2 Simultaneous desulfurization and denitrification experiment of 10%Fe_3O_4-titanium dioxide at different temperatures. Reprinted from Ref. [27], Copyright 2016, with permission from Elsevier

theory, i.e., it is a selective process of the material adsorbent. Based on the selective adsorption of molecular polarity, degree of unsaturation and the rate of polarizability, the molecular polarization is induced by electrostatic induction. So if the polarity is stronger or easier to be polarized, then the molecules are more likely to be adsorbed. Adsorbents such as Fe_3O_4-titanium dioxide have high adsorption capacity for both polar molecules and unsaturated molecules, and in nonpolar molecules, there is also a high selective adsorption capacity for the large polar molecules. So, there is a strong affinity of higher polarity such as H_2O and SO_2 molecules. And especially for water, it has a high adsorption capacity even under low pressure or low concentration and high temperature condition. While oxynitride is not highly polar gas [32, 33], the physical adsorption that occurs on the solid particles is relatively feeble. The similar phenomenon was also found in the study on simultaneous desulfurization and denitration. It may be due to the difference of molecular polarity and the adsorption dynamics of the two molecules, resulting in some differences in the absorption level of them. Generally, the SO_2 removal is prior to the removal of NO_x.

5.2.2.3 Analysis of Desulfurization and Denitrification Products of Fe_3O_4-Titanium Dioxide Composites

The 10% Fe_3O_4-titanium dioxide that had better desulfurization and denitration effect was used to make infrared spectrogram before and after the reaction (Fig. 5.3). As can be observed from Fig. 5.3, the differences of FTIR were large between the fresh and used sample, the large and broad band at 3387 cm^{-1} was attributed to the –OH stretching vibration on material surface, while the band at 1631 cm^{-1} was attributed to the –OH bending vibration band of H_2O. Compared with the unused adsorption materials, the band at 1047 and 1129 cm^{-1} corresponded to the S–O vibration band of SO_4^{2-} [34], which confirmed that the formation of SO_4^{2-} in the reaction made the

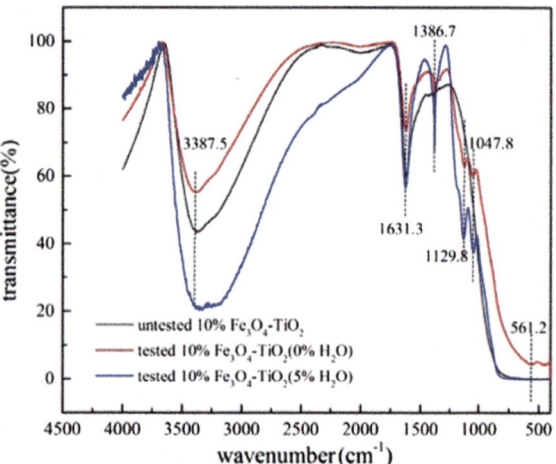

Fig. 5.3 FI-IR spectra of 10% Fe_3O_4-titanium dioxide composites before and after tests. Reprinted from Ref. [27], Copyright 2016, with permission from Elsevier

removal of SO_2 be achieved. Besides, the vibration peak in spectra of experiments with water was greater, which indicated that the addition of water can promote the formation of SO_4^{2-} [22]. A very sharp absorption peak was appeared at 1386 cm^{-1} of the used sample, which is the characteristic infrared absorption of nitrate [35, 36]. After adding 5% water, the vibration peak was even more sharp, which showed that the formation of nitrate increased. It could be that nitric acid was generated by the product of NO_2 after reacting with water [37]. This result indicated that NO_2 was formed by catalytic oxidation of NO and was adsorbed, further reacted with metal oxide on the carrier to form nitrate [38], achieving the removal of NO. Dalton etc. [39] demonstrated that after testing the surface reaction between the nitrogen oxide and surface of the titanium dioxide by XPS and Raman spectrometer, there was a good lattice defects on titanium dioxide surface, which had a good catalytic activity and can effectively convert nitrogen into nitrates.

The Fe_3O_4-titanium dioxide composites with stable structure were prepared by hydrothermal synthesis method. By the comparison of simultaneous desulfurization and denitration performance of Fe_3O_4-titanium dioxide, each single components and their physical mixture, it indicates the formation of a composite material, and the mutual promotion between the composite materials attribute to the best performance of simultaneous desulfurization and denitrification. It can be found that titanium dioxide was wrapped uniformly on the Fe_3O_4 surface by the characterization of XRD and TEM. The average size of the composite materials was in the range of 35–50 nm, the surface is the nano titanium dioxide crystalline with the size of nm 20–50 nm. The performance of 10% Fe_3O_4-titanium dioxide composite on simultaneous desulfurization and denitrification was tested at 100, 150 and 200 °C. The performance of the composite decreased gradually, and it had higher performance at the lower temperature. The removal efficiency of 10% Fe_3O_4-titanium dioxide composite on simultaneous desulfurization and denitrification was significantly higher than that of Fe_3O_4 and pure nano titanium dioxide. Through the FTIR characterization of Fe_3O_4-

titanium dioxide composite before and after reaction, it can be concluded that there are both physical adsorption and chemical adsorption in the reaction, being mainly chemical adsorption. SO_2 and NO_x are converted into sulfates and nitrates so that the removal of gas was achieved.

5.2.3 Mechanism of SO_2 Removal by TiO_2 Photocatalysis

It is reported by Luo [40] from their experiment about adsorption behavior of flue gas desulfurization using titanium dioxide that when spent titanium dioxide was desorbed by heating, 98% of SO_2 was escaped according to FTIR spectroscopy, demonstrating that the removal of SO_2 was mainly physical adsorption, resulting from Van der Waals force between SO_2 molecule and molecules absorbed on titanium dioxide surface [41, 42]. At the same time, it is also found that there was trace amount of SO_3 on spent titanium dioxide surface. This chemical absorption may be brought by chemical reaction between SO_2 molecules and molecules on titanium dioxide surface [43].

As for SO_2–O_2–N_2 system [44], SO_2 molecules produce single state ($^1SO_2$) and triplet state ($^3SO_2$) upon UV irradiation. The latter produces SO_3 by the reaction, $^3SO_2 + O_2 \rightarrow SO_3 + O$ directly, and there is deactivation of $^3SO_2$ [45]. The forming reaction of SO_3 does not proceed due to the absence of $^3SO_2$ without UV irradiation. Results from X-ray Photoelectron Spectroscopy (XPS) show that a broadband absorption ranging from 400–600 nm stands for the electron transition from the valence band to the conduction band, namely character band. Nevertheless, the response in 400–600 nm is related to surface property and ultramicro characteristic of titanium dioxide, which is called surface band [46]. According to experimental results of XPS, dissymmetry of O1s peak indicates there are different oxygen bonding states on the surface of titanium dioxide, i.e., the stronger peak in 529.5 eV belongs to O1s peak of Ti–O–Ti in the crystal lattice, and the weaker peak in 532.2 eV corresponds to O1s peak of the surface Ti–OH hydroxyl [47]. While reaction temperature increases, the quantity of surface hydroxyl might disappear, and the rate of photocatalytic oxidation declines. Combined the investigation with other results [40–47], the mechanism of SO_2 removal by titanium dioxide photocatalysis is inferred as follows:

$$\text{titanium dioxide} + h\nu \rightarrow e^- + h^+ \quad (5.3)$$

$$O_{2(g)} \rightarrow O_{2(ads)} \quad (5.4)$$

$$O_{(ads)} + e^- \rightarrow O^-_{(ads)} + h^+ \rightarrow O^*_{(ads)} \quad (5.5)$$

$$O^*_{(ads)} + H_2O_{(ads)} \rightarrow 2 \cdot OH_{(ads)} \quad (5.6)$$

$$\cdot OH_{(ads)} + SO_{2(ads)} \rightarrow HOSO_{2(ads)} \quad (5.7)$$

$$HOSO_{2(ads)} + O^*_{(ads)} \rightarrow \cdot OH_{(ads)} + SO_{3(ads)} \quad (5.8)$$

$$SO_{3(ads)} \rightarrow SO_{3(g)} \quad (5.9)$$

The key step of reaction rate is the reaction between $\cdot OH$ and SO_2 [46].

5.2.4 Mechanism of NO Removal by Titanium Dioxide Photocatalysis

Dalton et al. [39] studied the surface reaction of NO and titanium dioxide by Raman spectrometer and XPS. The results show that the anatase titanium dioxide has better crystalline defects and higher catalytic activity than rutile titanium dioxide, so NO can be converted by the former to nitrate efficiently. Experimental results reported by Hashimoto et al. [48] show that when titanium dioxide was irradiated in the presence of O_2, some active oxygen species such as super oxide (O_2^-) were generated, which reacted with NO_x to yield nitrate. Reaction equations are presumed as follows:

$$\text{titanium dioxide} + h\nu \rightarrow \text{titanium dioxide}^* \left(e_{cb}^- + h_{cb}^+ \right) \quad (5.10)$$

$$e_{cb}^- + O_{2(ads)} \rightarrow O_{2(ads)}^- \quad (5.11)$$

$$h_{cb}^+ + OH_{(ads)}^- \rightarrow \cdot OH_{(ads)} \quad (5.12)$$

$$NO_{(g)} + 2 \cdot OH \rightarrow NO_{2(ads)} + H_2O_{(ads)} \quad (5.13)$$

$$NO_{2(ads,g)} + \cdot OH \rightarrow NO_{3(ads)}^- + H_{(ads)}^+ \quad (5.14)$$

$$NO_{(ads)} + O_2^- \rightarrow NO_{3(ads)}^- \quad (5.15)$$

$$3NO_2 + 2OH- \rightarrow 2NO_3^- + NO + H_2O \quad (5.16)$$

$$[HNO_3]_{(ads)} \rightarrow HNO_{3(aq)} \quad (5.17)$$

5.3 Denitrification in Power Plant Flue Gas

5.3.1 The Principle of SCR to Remove NO_x

Governments have developed a corresponding emission reduction regulations and standards, in order to protect the ecological environment, reduce emissions, and so on. At present, the method of controlling the generation and emission of NO_x is mainly denitrification before combustion, denitrification in combustion and denitrification after combustion. Among the many denitrification technologies, SCR (selectively catalytic reduction) has many advantages such as no by-product, no secondary pollution, simple structure, reliable operation, easy maintenance and high denitrification efficiency, and it has already been widely used for commercial applications. Meanwhile it is the most widely used flue gas denitrification technology in this world [49]. In various De–NO_x technology, SCR is the most effective and mature one. In Japan, Shimoneski power plant built the first SCR system project in 1975. After that, this

5.3 Denitrification in Power Plant Flue Gas

technology was used widely through the country. So far, Japan, the US, China and many other countries have taken SCR as the main NO_x control technology. SCR has become a mature mainstream De–NO_x technology at home and abroad.

SCR is a post-burning NO_x control process. In an ammonia SCR system, the ammonia is sprayed into the coal boiler flue gas, and the mixed gas flows through a special reactor containing the catalyst. Under the action of catalysts, the ammonia reacts with NO_x to yield water and N_2. In the reaction process, NH_3 selectively reacts with NO_x instead of being oxidized [50]. So this reaction is also called selective reaction.

For SCR, the main reactions are shown in the following chemical formula.

$$4NH_3 + 4NO + O_2 \rightarrow 4N_2 + 6H_2O \quad (5.18)$$

$$8NH_3 + 6NO \rightarrow 7N_2 + 12H_2O \quad (5.19)$$

$$6NO + 4NH_3 \rightarrow 5N_2 + 6H_2O \quad (5.20)$$

$$2NO_2 + 4NH_3 + O_2 \rightarrow 3N_2 + 6H_2O \quad (5.21)$$

The efficiency of SCR depends to a great extent on the reaction activity of catalysts. Some other factors, such as the temperature, the reaction time, the mole ratio of NH_3 to NO_x, the flue gas flux, affect the efficiency very much. Generally speaking, the removal efficiency of NO_x in SCR is high enough to ensure that NH_3 and NO_x react nearly completely except that a small part of NH_3 escapes from the reactor. For fresh catalysts, ammonia escape is usually very low. But it will increase as the catalysts become deactivated and covered by dust on the surface [49, 51–55].

5.3.2 The Physical Shape Classification and Characteristics of SCR Catalyst

According to the division of processing mold and physical appearance, denitration catalysts are divided into three types, i.e., cellular, plate and crimped (triangle). [52].

(1) The cellular catalyst has Ti–W–V as the main active materials, which are fully mixed with titanium dioxide materials by mould extrusion and calcining. Its characteristics contain high catalytic activity in unit volume, so that it can achieve the high denitration efficiency with small volume of catalyst. It is suitable for the flue gas environment that ash content is less than 30 g/m and ash viscous is lower.

(2) The plate catalyst is with metal plate network as the skeleton and Ti–Mo–V as the main activity material, combining active material and the metal plate using the method of bilateral extrusion. Its module shape is similar to the heating plate of the air preheater. And it has a pitch of 6.0–7.0 mm and a smaller specific surface area. The characteristics of this type of catalyst is that it can be against strong corrosion and against blocking, suitable for the flue gas environment that

ash content and ash viscosity are high. However, it has disadvantages of low catalytic activity in unit volume, relatively high load, so it needs bigger volume and uses more steel structure.

(3) The crimped catalyst is made of glass fiber or has ceramic fiber as the skeleton. So its structure is quite solid. The aperture of this type of catalyst is relatively smaller. And it is close to cellular catalyst in unit volume efficiency. It also has a poor interchangeability with other types because of lower load, smaller volume of reactor and lower structure load. The crimped catalyst is always used in flue gas environment of low ash content [53].

5.3.3 The Chemical Material Classification of SCR Catalyst

According to the chemical material, the catalyst is classified into three types, i.e., platinum series, titanium series and vanadium series. At first, the platinum series is used widely. But nowadays, this type of catalyst is stopped using because of high price and high requirement of ash content. The most widely used catalyst is V_2O_5/WO_3, which uses titanium dioxide as the carrier and MoO_3. The main component of the catalyst is more than 99%. The rest trace components play important roles on the catalyst performance as well at the moment [53, 54].

5.3.4 High-Temperature and Low-Temperature Catalysts

The catalyst is classified into high-temperature and low-temperature types according to working temperature. High-temperature catalyst uses titanium dioxide, V_2O_5 as main ingredients. And its working temperature is 280–400 °C. So this type of catalyst is widely used in coal-fired power plants, fuel power plants and gas power plants. Low-temperature catalyst uses titanium dioxide, V_2O_5, MnO_2 as main ingredients. It has been used in fuel power plants and gas power plants because of the low working temperature of 180 °C. The low-temperature catalyst may be classified into four types: precious metal catalysts, molecular sieve catalyst, metal oxide catalyst and carbon materials catalysts.

5.3.4.1 Precious Metal Catalyst

The advantage of precious metal catalyst is that it has a better activity in low temperature. But it is expensive and easy to be oxygen inhibited and sulfur poisoned. This type of catalyst is made of some precious metal such as Pt, Rh and Pd and uses the alumina as the carrier. The earliest catalyst was precious metal catalyst. It was developed as the pollution control catalyst in 1970s. But it was instead by metal oxide catalyst in 1980s because of oxygen inhibition and sulfur poisoning. At present, this type of catalyst is only used in natural gas combustion flue gas and De–NO_x in low temperature. The research on this kind of catalyst focuses on further improving the selectivity of catalyst, sulfur resistance and low-temperature activity.

5.3.4.2 Molecular Sieve Catalyst

Molecular sieve catalyst is paid attention because it has a high activity and wide range of working temperature. The molecular sieve is an important factor that influences molecular sifter catalyst activity. Besides, the type of metal that exchanges ion with molecular sieve also affects the activity of molecular sieve catalyst [55, 56].

5.3.4.3 Metal Oxide Catalyst

Metal oxide catalyst is a mature kind of technology in the SCR technologies that are commonly used. Metal oxide is divided into some types, for instance, V_2O_5, Fe_2O_5, CuO, CrO, MnO, MgO and NiO. The catalysts that are researched and used most are V_2O_5/titanium dioxide, V_2O_5–WO_3/titanium dioxide and V_2O_5–WO_3/titanium dioxide. They can keep a high activity under the temperature of 300–400 °C. The De–NO_x activity of single metal oxide catalyst is not high, and it is not stable under high temperature. By stabilizing some active material, the activity of the composite metal oxide catalyst can be improved obviously when the composition and construction of composite metal oxide are under control [50, 57]. One common method is to load the oxide active component to the carrier by dipping. When the surface of the composite metal oxides gets activation treatment, it will have a higher thermal stability.

5.3.4.4 Carbon Materials Catalyst

The activated carbon has a microcrystalline structure, and the crystallites are completely irregularly arranged. The crystal has micropores, transition holes and large holes. It has a large inner surface and a specific surface area of 500–1700 m^2/g. This determines that the activated carbon has good adsorption properties and can adsorb metal ions, harmful gases, organic pollutants, pigments, etc. in waste water and exhaust gas. The application of activated carbon in the industry also requires high mechanical strength and good wear resistance. Its structure is stable and the energy required for adsorption is small to facilitate regeneration. Activated carbon is used for decolorization and deodorization of oils, beverages, foods, drinking water, gas separation, solvent recovery and air conditioning, and is used as a catalyst carrier and an adsorbent for gas masks. The catalyst activated carbon has a well-developed pore size and a large specific surface area, especially in the micro-mesopore volume.

On the whole, the SCR catalyst used in present projects contains unsupported metal oxide catalyst, metal oxide catalyst that uses titanium dioxide as the carrier and metal oxide catalyst that uses Al_2O_3, ZrO_2 and SiO_2 as the carrier. In these catalysts, the traditional metal oxide catalyst takes some special composite metal oxides as the active ingredients. These special composite metal oxides mainly use V_2O_5 as the primarily agent and use MoO_3, WO_3 and MoO_3 as auxiliary agent [58, 59].

References

1. S.A. Guerra, S.R. Olsen, J.J. Anderson, Evaluation of the SO_2 and NO_x offset ratio method to account for secondary PM2.5 formation. J. Air Waste Manag. Assoc. **64**, 265–271 (2014)
2. L. Tang, T. Nagashima, K. Hasegawa, T. Ohara, K. Sudo, N. Itsubo, Development of human health damage factors for PM2.5 based on a global chemical transport model. Int. J. Life Cycle Assess., 1–11 (2015)
3. I. Mochida, Y. Korai, M. Shirahama, S. Kawano, T. Hada, Y. Seo, M. Yoshikawa, A. Yasutake, Removal of SO_x and NO_x over activated carbon fibers. Carbon **38**, 227–239 (2000)
4. Y.T. Li, Y.J. Mao, Q. Zhong, H.X. Qu, J. Wang, Effects of components of SCR catalyst on NO_x performance. J. Fuel Chem. Technol. **37**, 601–606 (2009)
5. Y. Zhao, Z.G. Han, Y.H. Han, J. Yao, Application and new progress of dry flue gas simultaneous desulfurization and denitrification technology. Ind. Saf. Environ. Prot. **4**, 4–6 (2009)
6. Y. Gao, H. Liu, Preparation and catalytic property study of a novel kind of suspended photocatalyst of titanium dioxide-activated carbon immobilized on silicone rubber film. Mater. Chem. Phys. **92**, 604–608 (2005)
7. H. Yamashita, M. Harada, J. Misaka, M. Takeuchi, K. Ikeue, M. Anpo, Degradation of propanol diluted in water under visible light irradiation using metal ion-implanted titanium dioxide photocatalysts. J. Photochem. Photobiol., A **148**, 257–261 (2002)
8. X. Zhao, L. Lv, B. Pan, W. Zhang, S.J. Zhang, Q.X. Zhang, Polymer-supported nanocomposites for environmental application: a review. Chem. Eng. J. **170**, 381–394 (2011)
9. S.K. Kim, H. Chang, K.K. Cho, D.S. Kila, S.W. Cho, H.D. Jang, J.W. Choi, J. Choi, Enhanced photocatalytic property of nanoporous titanium dioxide/SiO_2 micro-particles prepared by aerosol assisted co-assembly of nanoparticles. Mater. Lett. **65**, 3330–3332 (2011)
10. L.H. Mahajan, S.T. Mhaske, Composite microspheres of poly (o-anisidine)/titanium dioxide. Mater. Lett. **68**, 183–186 (2012)
11. M. Kapilashrami, Y. Zhang, Y-S Liu, A. Hagfeldt, J. Guo, Probing the optical property and electronic structure of TiO_2 nanomaterials for renewable energy applications. Chem. Rev. **114**(19), 9662–707 (2014)
12. Y. Liu, L. Wang, W. Jin, C. Zhang, M. Zhou, W. Chen, Synthesis and photocatalytic property of $TiO_2@V_2O_5$ core-shell hollow porous microspheres towards gaseous benzene. J. Alloy. Compd. **690**, 604–11 (2017)
13. Y. Ma, X. Wang, Y. Jia, X. Chen, H. Han, C. Li, Titanium dioxide-based nanomaterials for photocatalytic fuel generations. Chem. Rev. **114**(19), 9987–10043 (2014)
14. H. An, L. Pan, H. Cui, D. Zhou, B. Wang, J. Zhai, et al., Electrocatalytic performance of Pd nanoparticles supported on TiO_2-MWCNTs for methanol, ethanol, and isopropanol in alkaline media. J. Electroanal. Chem. **741**, 56–63 (2015)
15. R. Zhou, J. Wu, J. Zhang, H. Tian, P. Liang, T. Zeng, et al., Photocatalytic oxidation of gas-phase Hg^0 on the exposed reactive facets of BiOI/$BiOIO_3$ heterostructures. Appl. Catal. B-Environ. **204**, 465–74 (2017)
16. W.-J. Ong, L.-L. Tan, Y.H. Ng, S.-T. Yong, S.-P. Chai, Graphitic carbon nitride (g-C_3N_4)-based photocatalysts for artificial photosynthesis and environmental remediation: are we a step closer to achieving sustainability? Chem. Rev. **116**(12), 7159–329 (2016)
17. L. Ding, R. Wei, H. Chen, J. Hu, J. Li, Controllable synthesis of highly active BiOCl hierarchical microsphere self-assembled by nanosheets with tunable thickness. Appl. Catal. B-Environ. **172–173**, 91–9 (2015)

18. S. Sarina, E.R. Waclawik, H. Zhu, Photocatalysis on supported gold and silver nanoparticles under ultraviolet and visible light irradiation. Green Chem. **15**(7), 1814–33 (2013)
19. X. Chen, S.S. Mao, Titanium dioxide nanomaterials: synthesis, properties, modifications, and applications. Chem. Rev. **107**(7), 2891–959 (2007)
20. K. Li, B. Peng, T. Peng, Recent advances in heterogeneous photocatalytic CO_2 Conversion to solar fuels. ACS Cat. **6**(11), 7485–527 (2016)
21. X.M. Qi, M.L. Gu, X.Y. Zhu, J. Wu, H.M. Long, K. He, et al., Fabrication of $BiOIO_3$ nanosheets with remarkable photocatalytic oxidation removal for gaseous elemental mercury. Chem. Eng. J. **285**, 11–9 (2016)
22. L.P. Belo, L.K. Elliott, R.J. Stange, R. Spörl, K.V. Shah, J. Maier, T.F. Wall, Hightemperature conversion of SO_2 to SO_3: homogeneous experiments and catalytic effect of fly ash from air and oxy-fuel firing. Energy Fuels **28**, 7243–7251 (2014)
23. J. Gao, R.Z. Jiang, J. Wang, P.L. Kang, B.X. Wang, Y. Li, K. Li, Z.D. Zhang, The investigation of sonocatalytic activity of Er^{3+}: $YAlO_3$/titanium dioxide-ZnO composite in azo dyes degradation. Ultrason. Sonochem. **18**, 541–548 (2011)
24. Y. Liu, L. Yu, Y. Hu, C.F. Guo, F.M. Zhang, X.W. (David) Lou, A magnetically separable photocatalyst based on nest-like c-Fe_2O_3/ZnO double-shelled hollow structures with enhanced photocatalytic activity. Nanoscale **4**, 183–187 (2012)
25. M. Ye, Q. Zhang, Y. Hu, J. Ge, Z. Lu, L. He, Z.L. Chen, Y.D. Yin, Magnetically recoverable core-shell nanocomposites with enhanced photocatalytic activity. Chem. A Eur. J. **16**, 6243–6250 (2010)
26. G. Cheng, Z.G. Wang, Y.L. Liu, J.L. Zhang, D.H. Sun, J.Z. Ni, Magnetic affinity microspheres with meso-/macroporous shells for selective enrichment and fast separation of phosphorylated biomolecules. ACS Appl. Mater. Interfaces. **5**, 3182–3190 (2013)
27. Y. Li, H. Yi, X. Tang et al., Study on the performance of simultaneous desulfurization and denitrification of Fe_3O_4-titanium dioxide composites. Chem. Eng. J. **304**, 89–97 (2016)
28. F. Delbecq, P. Sautet, Interplay between magnetism and chemisorption: a theoretical study of CO and NO adsorption on a Pd3Mn alloy surface. Chem. Phys. Lett. **302**, 91–97 (1999)
29. S. Bennici, G. Antonella, M. Lazzarin, V. Ragaini, CuO based catalysts on modified acidic silica supports tested in the de-NO_x reduction. Ultrason. Sonochem. **12**, 307–312 (2005)
30. H. Ichiura, T. Kitaoka, H. Tanaka, Photocatalutic oxidation of NO_x using composite sheets containing titanium dioxide and a metal compound. Chemosphere **51**, 855–860 (2003)
31. L. Zhao, Experimental study on photocatalytic oxidation for simultaneous desulfurization and denitrification, North China Electric Power University, 2007
32. E.S. Kikkinides, R.T. Yang, Gas separation and purification of polymeric adsorbents. Ind. Eng. Chem. Res. **32**, 4063–4077 (1995)
33. J. Villadsen, M.L. Micheisen, Solution of differential equation models by polynomial approximation. AIChE J. **25**, 1345–1359 (1994)
34. Q. He, Z.X. Zhang, J.W. Xiong, Y.Y. Xiong, H. Xiao, A novel biomaterial-Fe_3O_4: titanium dioxide core-shell nano particle with magnetic performance and high visible light photocatalytic activity. Opt. Mater. **31**, 380–384 (2008)
35. M. Kantcheva, A.S. Vakkasoglu, Cobalt supported on zirconia and sulfated zirconia. I: FT-IR spectroscopic characterization of the NO_x species formed upon NO adsorption and NO/O_2 coadsorption. J. Catal. **223**, 352–363 (2004)
36. Q. Wang, S.Y. Park, J.S. Choi, J.S. Chung, $Co/K_xTi_2O_5$ catalysts prepared by ion exchange method for NO oxidation to NO_2. Appl. Catal. B **79**, 101–107 (2008)
37. J.X. Zhang, Studies on absorption of SO_2 and NO_x by activated carbon, Xi'an University of Architecture and Technology, 2008
38. H.Y. Huang, R.T. Yang, Removal of NO by reversible adsorption on Fe–Mn based transition metal oxides. Langmuir **17**, 4997–5003 (2001)
39. J.S. Dalton, P.A. Janes, N.G. Jones, J.A. Nicholson, K.R. Hallam, G.C. Allen, Photocatalytic oxidation of NO_x gases using titanium dioxide: a surface spectro-scopic approach. Environ. Pollut. **120**, 415–422 (2002)

40. Y.Q. Luo, D.J. Li, Z. Huang, Preparation of titanium dioxide particles and properties for flue gas desulfurization. Environ. Sci. **24**(1), 147–151 (2003). (in Chinese)
41. R.T. Yang, W.B. Li, N. Chen, Reversible chemisorption of nitric oxide in the presence of oxygen on titania and titania modified with surface sulfate. Appl. Catal. A-Gen. **169**, 215–225 (1998)
42. Y.S. Lin, S.G. Deng, Analysis of liquid chromatography with nonuniform crystallite particles. AIChE J. **36**(10), 1569–1576 (1990)
43. Y. Wang, S.O. Mohammed, J.C. Lavelley et al., FTIR study of adsorption and reaction of SO_2 and H_2S on Na/SiO_2. Appl. Catal. B-Environ. **16**, 279–290 (1998)
44. J. Shang, Z.L. Xu, G.Y. Du et al., Studies on photocatalytic oxidation reaction SO_2 over titanium dioxide. Chem. J. Chin. U **21**, 1299–1300 (2000)
45. X.Y. Tang, *Atmosphere environmental chemistry* (High Education Press, Beijing, 1990), pp. 124–125
46. R. Campostrini, G. Carturan, L. Palmisano, Sol-gel derived anatase titanium dioxide morphology and photoactivity. Mater. Chem. Phys. **38**(3), 277–283 (1994)
47. J. Sanchez, J. Augustynski, X-ray photoelectron spectroscopic study of the interaction of various anions with the oxide-covered titanium metal. J. Electroanal. Chem. **103**(26), 423–426 (1979)
48. K. Hashimoto, K. Wasada, N. Toukai et al., Photocatalytic oxidation of nitrogen monoxide over titanium (VI) oxide nanocrystals large size areas. J. Photochem. Photobiol., A **136**, 103–109 (2000)
49. M. Romero, J. Blanco, B. Sanchez et al., Solar photocatalytic degradation of water and air pollutants: challenges and perspectives. Sol. Energy **66**, 169–182 (1999)
50. Yi Zhao, Hongtao Zhu, Xiaoling An, Su Peng, Study on SCR technique for flue gas denitrification in coal-fired power plants. Electr. Power Environ. Prot. **25**(1), 7–10 (2009). (In Chinese)
51. L. Sun, X. Li, Applications of SCR denitrification technology in coal-fired power plants. Sci. Technol. Assoc. Forum, **2**(2), 37 (2010)
52. H. Chen, X. Song, H. Jiang, Z. Cui, G. Zhang, Maintenance and affecting factors for the performance of SCR system. J. Shandong Jianzhu Univ. **23**(2), 145–149 (2008). (In Chinese)
53. S. Wang, X. Huang, The development of catalysts for selective catalytic reduction of NO_X in flue gas. China Chem. **5**, 55–60 (2009). (In Chinese)
54. Y. Wang, Y. Sun, F. Chen, Y. Lin, X. Liang, Characteristics of SCR catalyst and its application in coal-fired power plants De-NO_x system. Electr. Power Environ. Prot. **25**(4), 13–15 (2009). (In Chinese)
55. Kang Liu, Qiang Zhang, Yu. Hong, Yu. Hong, Species and application of SCR catalyst for flue gas denitrification in coal-fired power plant. Electr. Power Environ. Prot. **25**(4), 9–12 (2009). (In Chinese)
56. Xu Furong, Lirong Zhou, Discussion about processing scheme for disabled SCR catalyst in coal-fired power plant. China Environ. Projection Ind. **11**, 25–27 (2010). (In Chinese)
57. F. Zhao, M.C. Lai, D.L. Harrington, Automotive spark-ignited direct injection gasoline engines. Prog. Energy Combust. Sci. **25**, 437–562 (1999)
58. J.W. Hosch, J.P. Walters, High spatial resolution schlieren photography. Appl. Opt. **16**(2), 473–485 (1977)
59. J.H. Li, J.M. Hao, L.X. Fu et al., The activity and characterization of sol-gel Sn/Al_2O_3 catalyst for selective catalytic reduction of NO_X in the presence of oxygen. Catal. Today **90**, 215–221 (2004)

Chapter 6
The Photocatalytic Removal of Mercury from Coal-Fired Flue Gas

Abstract Mercury in the flue gas of coal combustion exists in three chemical forms: elemental mercury (Hg^0), particle-bounded mercury (Hg^p), and oxidized mercury (Hg^{2+}). Hg^{2+} is water-soluble, so it can be removed by wet scrubbers and the Hg^{2+} removal efficiency can reach up to 90%. In addition, Hg^p can be easily removed by electrostatic precipitator (ESP) or fabric filter (FF). However, elemental mercury is difficult to remove because of its volatility, insolubility and chemical stability. So the most critical step is to control elemental mercury. There are two main methods for elemental mercury control: adsorption and oxidation. Activated carbon injection has proven to be an effective adsorption method for elemental mercury removal in flue gas. However, its high operation cost and negative effect on fly ash quality may restrict its industrial application. The drawbacks associated with the use of particulate adsorbents make the oxidation method more attractive. In recent years, the progress in the photocatalytic oxidation of elemental mercury has aroused wide interests among researchers. Compared with other methods, photocatalytic oxidation possesses higher oxidation ability and no secondary pollution, therefore it is a promising technology for oxidation of elemental mercury. We set up a chemical reactor system to evaluate the photocatalytic performance on mercury in the simulated flue gas, and studied the mercury oxidation efficiency of different photocatalysts, including morphology controlled photocatalysts, metal or nonmetal modified titanium dioxide photocatalysts and others photocatalysts, such as $BiOIO_3$ and ZnO. The morphology controlled photocatalysts include hollow titanium dioxide photocatalyst and anatase titanium dioxide with co-exposed (001) and (101) facets. The metal modified titanium dioxide photocatalysts include CuO/titanium dioxide and V_2O_5/titanium dioxide photocatalysts, and the nonmetal modified titanium dioxide photocatalysts include carbon spheres supported CuO/titanium dioxide photocatalysts and carbon decorated In_2O_3/titanium dioxide photocatalysts. These studies are helpful to observe the characteristics and performance of different kinds of photocatalysts.

Keywords Homojunction · Heterojunction · Redox pair · Photocatalytic Mercury

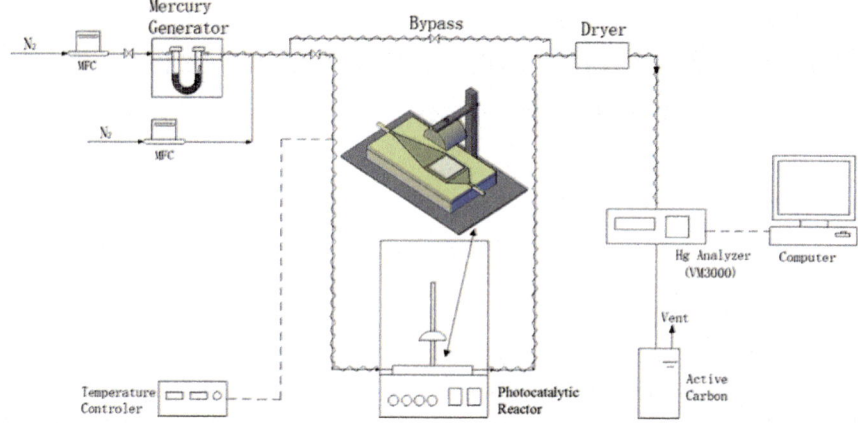

Fig. 6.1 Schematic diagram of the experimental system. Reprinted from Ref. [1], Copyright 2015, with permission from Elsevier

6.1 Measurement of Photoactivity

The schematic diagram of the experimental system was shown in Fig. 6.1. The system consisted of simulated flue gas, a photocatalytic reactor and a mercury analyzer. The N_2 was divided into two branches, whose flow rates were controlled by two mass flow meters (MFC, CS200 type) separately. The total flow of the two branches was 1.2 L/min. One gas stream with a flow rate of 0.2 L/min passed through the elemental mercury permeation tube to introduce elemental mercury vapor into the system. The elemental mercury inlet concentration was around 65 $\mu g/m^3$. The mercury permeation tube was placed in a U-shape glass tube which was immersed in a constant 55 °C water bath to ensure a constant elemental mercury permeation rate. The other stream of N_2 had a flow rate of 1L/min. An on-line mercury analyzer (VM3000 mercury vapor monitor, Mercury Instruments, Germany) based on atomic absorption spectrometry was used to measure the gas phase concentration of elemental mercury. The gas flowed through two silica gel columns before entering the mercury analyzer to remove water vapor. Finally, the gas got through the active carbon bottle and vented out of the system.

The catalyst (50 mg) was coated on quartz glass plates (75 mm × 75 mm) by using a dip-coating method, and then it was put into the photocatalytic reactor developed by ourselves, which was shown in Fig. 6.2. The photocatalytic reactor was consisted of inlet transition zone, reaction zone and outlet transition zone.

The reaction zone had quartz glass cover plate on top, so light could reach the surface of the catalyst through the cover board. Reflective device was placed around the tubes, making the illumination uniform throughout the reaction zone. The simulated flue gas containing mercury went through the inlet transition zone between the inlet and the reaction zone and accepted the light to complete the photocatalytic

6.1 Measurement of Photoactivity

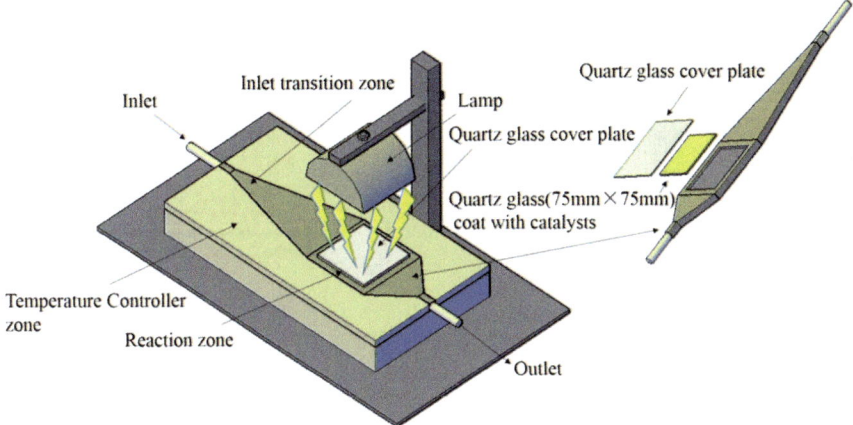

Fig. 6.2 Schematic of the photocatalytic reactor. Reprinted from Ref. [1], Copyright 2015, with permission from Elsevier

oxidization process. The inlet transition zone made the flow field in the reaction zone uniform so that the reaction process would not be distorted by nonuniform flow field and better simulate the flow field in the flue gas duct in coal-fired power plants.

In this work, in order to describe this phenomenon in detail, the elemental mercury removal efficiency under light irradiation condition was defined as following:

$$\eta = \frac{Hg_{in}^0 - Hg_{out}^0}{Hg_{in}^0} \times 100\% \qquad (6.1)$$

where Hg_{in}^0 represented $Hg^0(\mu g)$ at the inlet of the reactor, and Hg_{out}^0 represented $Hg^0(\mu g)$ at the outlet of the reactor.

6.2 The Photocatalytic Removal of Mercury by Metal or Nonmetal Modified Titanium Dioxide Photocatalysts

6.2.1 The Photocatalytic Removal of Mercury by V_2O_5/Titanium Dioxide Photocatalysts

6.2.1.1 Effect of Calcination Temperature

The photocatalytic activity of synthesized nanocomposites were examined under visible light. The 24 W LED light was used as the visible light source, and its wavelength was 420 nm. Each test was performed for around 150 min, and the photocatalytic oxidation of mercury was assessed under dark (30 min) and visible

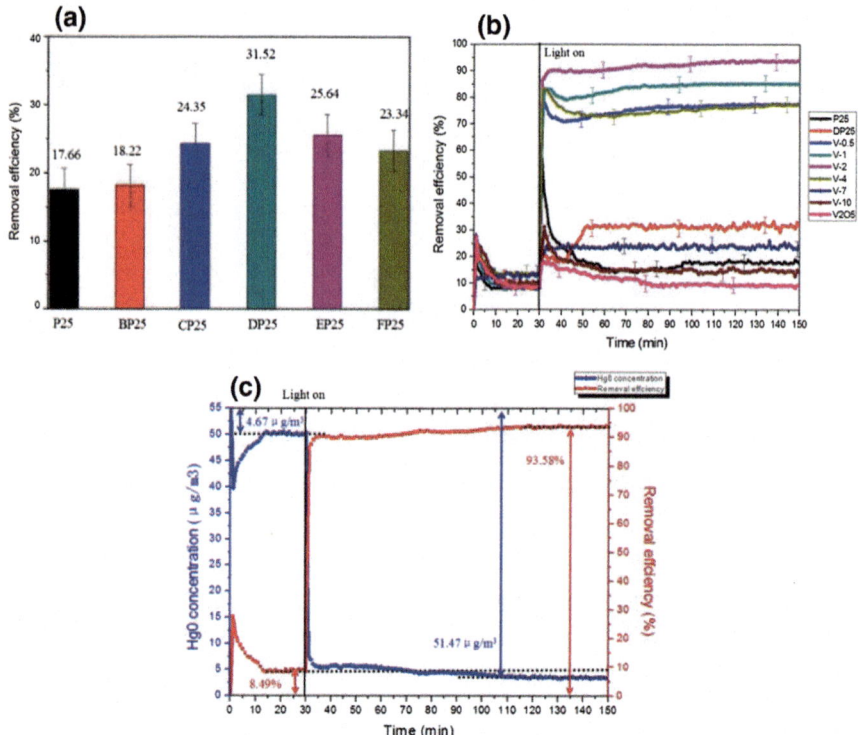

Fig. 6.3 **a** Hg^0 photocatalytic oxidation efficiencies of as-prepared nanocomposites with (a) calcination temperature; **b** vanadium doping amounts; **c** The changes of Hg^0 concentrations and oxidation efficiency for V-2 with time. Reprinted from Ref. [3], Copyright 2017, with permission from Elsevier

light irradiation (120 min) respectively. The influence of calcination temperature on photocatalytic performance was measured. As shown in Fig. 6.3a, as the calcination temperature increases, the photocatalytic oxidation efficiency increases. However, when the calcination temperature exceeds 400 °C, the opposite phenomenon occurs. This indicates that the samples with calcination temperature 400 °C show the highest photocatalytic performance.

6.2.1.2 Preparation of V_2O_5/Rutile-Anatase Photocatalyst System

V_2O_5/rutile-anatase photocatalyst system with different V/Ti element molar ratios ranging from 0:1 to 0.1:1 was prepared by using incipient wet impregnation method. In a typical synthesis (V/Ti element molar ratio is 0.02:1), ammonium metavanadate (NH_4VO_3, 0.0585 g, 0.5 mmol) was added into 60 mL of hot deionized water and stirred continuously for 15 min. Then P25 TiO_2 (1.9968 g, 25 mmol) was added to the solution with continuous stirring for 30 min and dried at 120 °C in a ventilation oven overnight. Then the dry samples were ground in an agate mortar with pestle and the obtained powder calcined in a muffle furnace. The muffle furnace temperature was 400 °C with the heating rate of 5 °C/min and held for 3 h. 0, 0.5, 1, 2, 4, 7, and 10 mol% of V-doped rutile-anatase nanocomposites hereafter named as DP25, V-0.5, V-1, V-2, V-4, V-7 and V-10 respectively.

6.2.1.3 Effect of V_2O_5 Content

In Fig. 6.3b, the effect of different V doping ratios (from 0 to 10 mol% of V/Ti) on the oxidation of Hg^0 was investigated. As can be seen that the oxidation efficiency of DP25 is higher than that of P25, and oxidation efficiency of P25 and DP25 are 17.66 and 31.52% respectively. This is mainly due to the fact that homojunction between anatase titanium dioxide and rutile titanium dioxide play a very important role in process of photocatalytic oxidation. In addition, DP25 has larger specific surface area than P25. It means that DP25 expose more active sites, which is benefit to photocatalytic oxidation of Hg^0. Futhermore, the Hg^0 oxidation efficiency over as-prepared V_2O_5/rutile-anatase nanocomposites increased as the content of V_2O_5 increased, and then decreased as the V_2O_5 loaded further increased. Therefore, the optimal molar ratio of V_2O_5/rutile-anatase is 2 mol%, and the highest photocatalytic efficiency is about 93.58%, which is much higher than that of P25 under visible light. As shown in Fig. 6.3c, mercury concentration was initially about 55 $\mu g/m^3$, which decreased 4.67 and 51.47 $\mu g/m^3$ respectively, corresponding to oxidation efficiencies of 8.49 and 93.58% respectively in dark and visible light. The reason for the observed higher photocatalytic activity for the as-prepared V_2O_5/rutile-anatase nanocomposites samples can be due to the formation of homo-hetero junctions among the vanadia species, rutile titanium dioxide, and anatase titanium dioxide. Photogenerated electrons are moved on the conduction band (CB), which commence at V_2O_5 and gather on the CB of anatase. Photogenerated holes migrate on the valence band (VB), which start with anatase and assemble on the CB of V_2O_5. In this process, it is possible to suppress the recombination of photogenerated electrons and holes and promote separation and migration of electrons and holes at the interfaces and help to produce ·OH, which is set as the active components for promoting photocatalytic performance [2]. Excessive V_2O_5 contents are detrimental to the photocatalytic activity, because the excessive V_2O_5 particles may cover active sites on as-prepared samples surface, reducing the photocatalytic activity.

6.2.1.4 The Stability of As-Prepared Photocatalyst

The cyclic stability is an important factor for practical application of photocatalyst. Cyclically photocatalytic oxidation of Hg^0 was conducted to examine the stability of as-prepared samples. The first and second groups were carried out for about 120 min and the third group was irradiated under visible light (24 W LED) for about 160 min (24 W LED), and the results are shown in Fig. 6.4a. In the test, the quartz glass loaded with V-2 photocatalyst was initially detected under dark for 30 min and visible light was turned on hereafter. After performing the experiment for 120 min, the visible light was turned off and the next cyclic experiment was conducted after the Hg^0 concentration stabilized for a while. The same procedures were carried out in the second and third experiments, and the only difference was that the third time was performed for 160 min. It was shown that the oxidation efficiency of Hg^0 hardly changed in three cycles with long time, and the V-2 sample exhibits remarkable and stable photocatalytic activity for removing Hg^0. The structures of the V-2 sample before and after three times of consecutive cycle experiments over a long period are also characterized by the XRD and XPS analysis, as exhibited in Fig. 6.4b, c, which show no observable change after this reaction. The result shows that V-2 sample possesses higher photocatalytic activity and chemical structure stability even when the cycle test is repeated for a long time.

6.2.1.5 Effect of Individual Flue Gas Components

By comparing oxidation efficiencies of Hg^0 by V-2 sample in pure N_2 and mixture atmosphere of N_2 and other individual flue gas components, Fig. 6.5 shows the effects of the individual flue gas component on the Hg^0 oxidation efficiency by V-2 sample. In the coal-burning power plants, sulfur-tolerance is an important factor for catalyst owing to the fact that SO_2 is one of main components in flue gases. From Fig. 6.5a, it can be seen that SO_2 has an adverse effect on the Hg^0 oxidation and the oxidation efficiency dropped from 42.72 to 36.93% as the SO_2 concentration increased from 200 to 400 ppm. The reason why the Hg^0 oxidation decreased with the SO_2 concentration increasing may be that a weak interaction being existed between SO_2 and V_2O_5 resulted in SO_2 concentration on the surface of V-2 sample increasing shown as Fig. 6.5b and in turn reduced the amount of active sites for Hg^0 adsorption.

The influence of NO on the oxidation efficiency of Hg^0 are shown in Fig. 6.5c. The inhibition of Hg^0 oxidation by NO also occurs for V-2 sample and reduces the oxidation efficiency from 93.58 to 54.73% (100 ppm NO) and 49.05% (300 ppm NO) respectively. The role of NO in Hg^0 oxidation over V-2 sample can be summarized as follows shown in Fig. 6.5d. First, the gas-phase NO may directly react with Hg^0 adsorbed on the surface of as-prepared sample. However, only a small amount of NO react in this way. Second, many NO molecules are adsorbed onto the surface of V-2 sample, which hinder the adsorption of Hg^0. According to above analysis, with NO concentration increasing, the photocatalytic efficiency of sample tends to decrease. Third, active nitrogen component, for instance, NO_2 is formed onto the surface because NO can react with O_2 and $\cdot O_2^-$, which are more beneficial for oxidizing and adsorbing of Hg^0 than NO [3, 4]. Fourth, reactive components ($\cdot OH$, $\cdot O_2^-$, and h^+) react with NO, decreasing the reaction between reactive species and Hg^0. The limitation mainly includes the following reaction equations [5]:

$$NO(g) \rightarrow NO(ad) \tag{6.2}$$

6.2 The Photocatalytic Removal of Mercury by Metal or Nonmetal ...

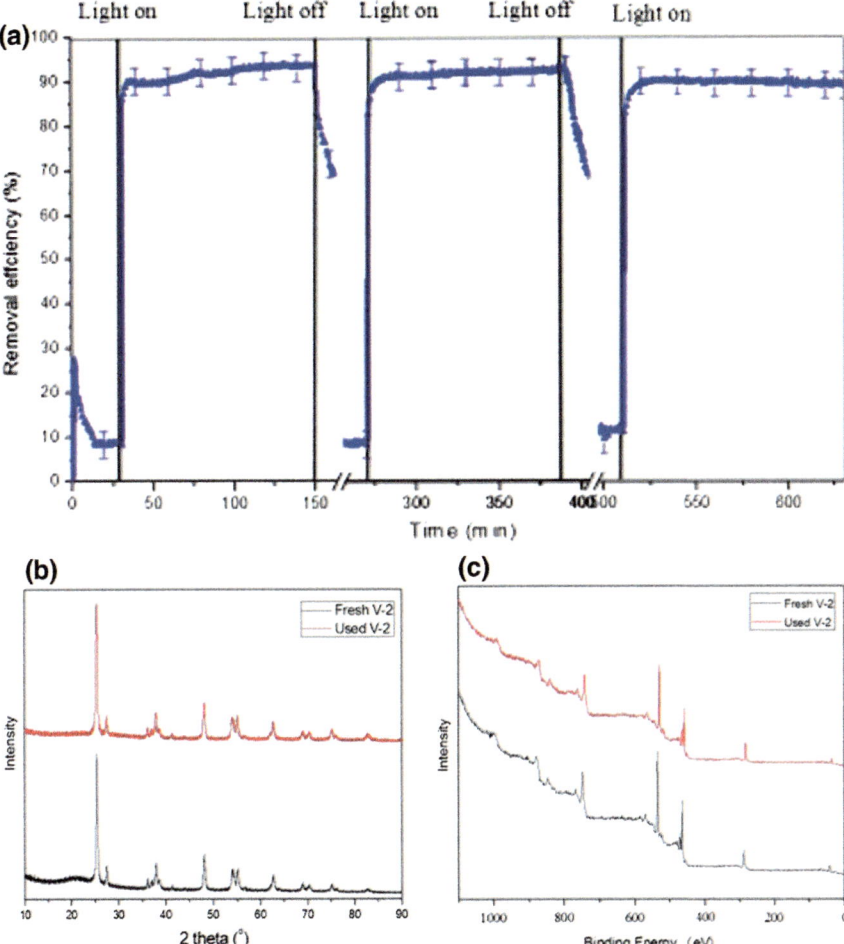

Fig. 6.4 **a** Stability evaluation for the V-2 sample after three consecutive cycling photocatalytic oxidation of Hg^0; **b** XRD patterns of V-2 sample before and after recycling reactions; **c** XPS of V-2 sample before and after recycling reactions. Reprinted from Ref. [3], Copyright 2017, with permission from Elsevier

$$Hg^0(g) \rightarrow Hg^0(ad) \tag{6.3}$$

$$2NO(ad) + 2Hg^0(ad) \rightarrow N_2(g) + 2HgO(ad) \tag{6.4}$$

$$2NO(ad) + O_2(g) \rightarrow 2NO_2(ad) \tag{6.5}$$

$$2NO(ad) + \cdot O_2^- \rightarrow 2NO_2(ad) \tag{6.6}$$

$$NO_2(ad) + Hg^0(ad) \rightarrow HgO(ad) + NO(g) \tag{6.7}$$

$$NO(ad) + 3 \cdot OH(ad) \rightarrow NO_3^- + H^+ + H_2O \tag{6.8}$$

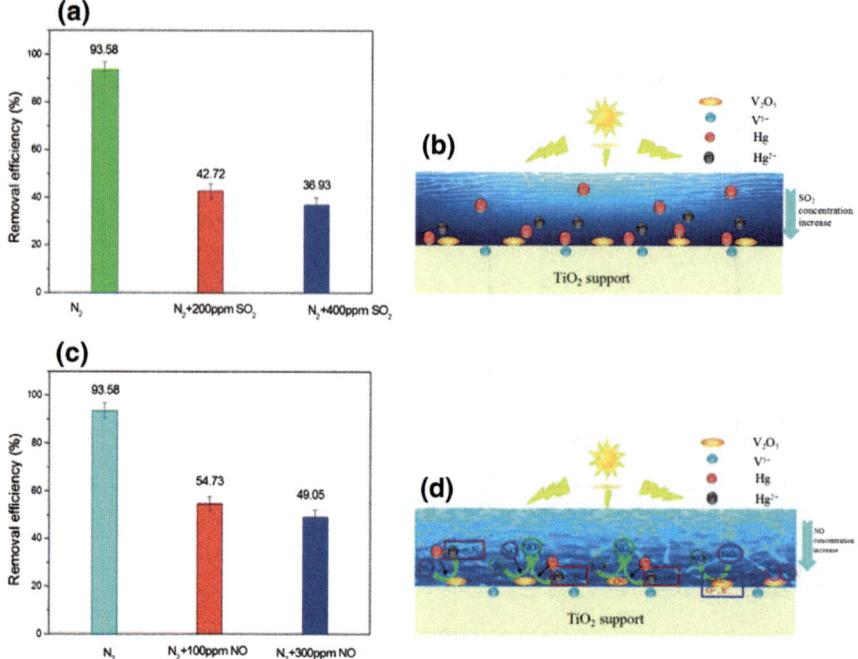

Fig. 6.5 a Effect of SO$_2$ on Hg^0 oxidation efficiency; **b** The possible mechanism of SO$_2$ effect on Hg^0 oxidation; **c** Effect of NO on Hg^0 oxidation efficiency; **d** The possible mechanism of NO effect on Hg^0 oxidation. Reprinted from Ref. [3], Copyright 2017, with permission from Elsevier

$$NO(ad) + \cdot O_2^-(ad) \rightarrow NO_3^- \tag{6.9}$$

$$2NO(ad) + 2h^+ + O_2 + 2H_2O \rightarrow 2NO_3^- + 4H^+ \tag{6.10}$$

6.2.1.6 Possible Photocatalytic Mechanism

Figure 6.6a–c schematically shows the proposed reaction mechanism in V$_2$O$_5$/rutile-anatase photocatalyst system for Hg^0 oxidation. The as-prepared V$_2$O$_5$/rutile-anatase photocatalysts show surpassing Hg^0 oxidation efficiency, which is much higher than that of pure P25 under visible light irradiation. The potential of the CB boundary of the semiconductors is defined by Mott-Schottky (M-S) curve [6, 7]. As shown in Fig. 6.6d, the inclination of M-S plot is positive, which reveals that the as-prepared V$_2$O$_5$ is n-type semiconductor. The CB potential can be determined as the x-intercept of M-S plot. The calculated CB potential is 0.19 eV. Theoretically, the band gaps of V$_2$O$_5$, anatase and rutile are 2.24, 3.20 and 3.02 eV respectively, and the CB potentials of anatase and rutile are −0.18 and −0.37 eV respectively, shown in Fig. 6.6e [8]. After V$_2$O$_5$, anatase, and rutile contacting, ternary nanocomposites have the same Fermi energy level, which leads that the conduction band (CB) and valence band

6.2 The Photocatalytic Removal of Mercury by Metal or Nonmetal … 111

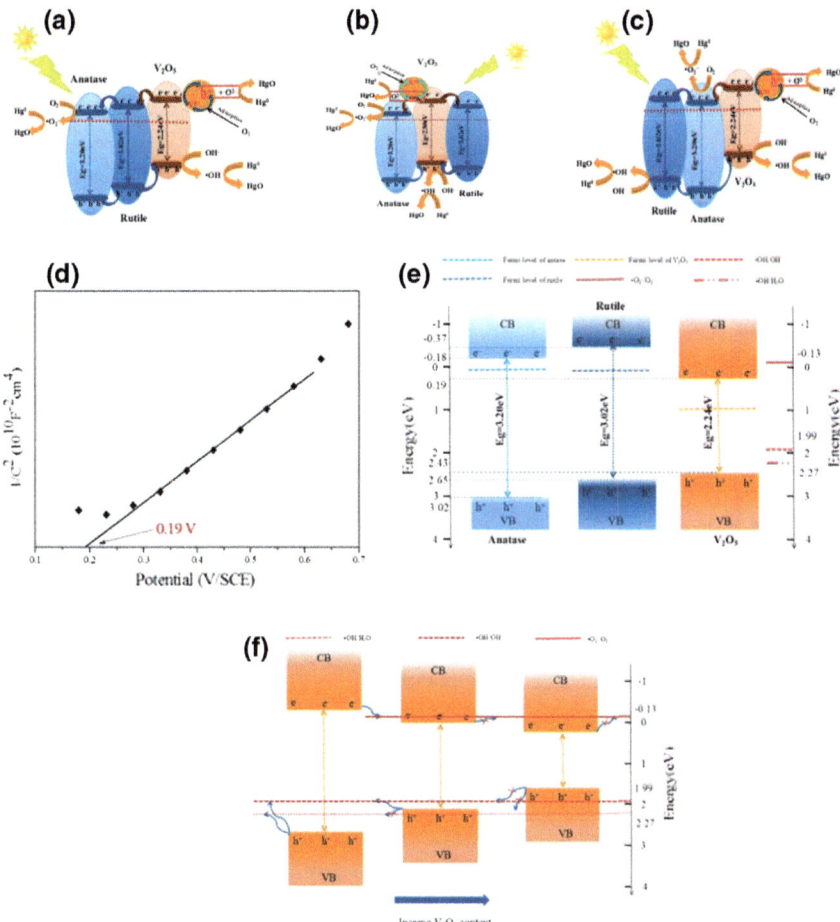

Fig. 6.6 **a** Schematic illustration for the theory one to charge transfer in the V_2O_5/rutile-anatase photocatalyst system; **b** Schematic illustration for the theory two to charge transfer in the V_2O_5/rutile-anatase photocatalyst system; **c** Schematic illustration for the theory three to charge transfer in the V_2O_5/rutile-anatase photocatalyst system; **d** Mott-Schottky plot of pure V_2O_5; **e** The conduction band (CB) and valence band (VB) of anatase, rutile, and V_2O_5 before contraction; **f** Impact of V_2O_5 on the CB and VB level of V_2O_5/rutile-anatase photocatalyst system; **g** O 1s XPS spectra of the used V-2; **h** V 2p XPS spectra of the used V-2; **i** Hg 4f XPS spectra of the used V-2. Reprinted from Ref. [3], Copyright 2017, with permission from Elsevier

(VB) of V_2O_5 lie over that of rutile while the CB and VB of rutile lie over the energy band of anatase respectively [9]. According to the previous report, the CB of anatase lies 0.2 eV under the CB of rutile [10]. Under visible light irradiation, V_2O_5/rutile-anatase photocatalyst system can simultaneously generate electrons and holes.

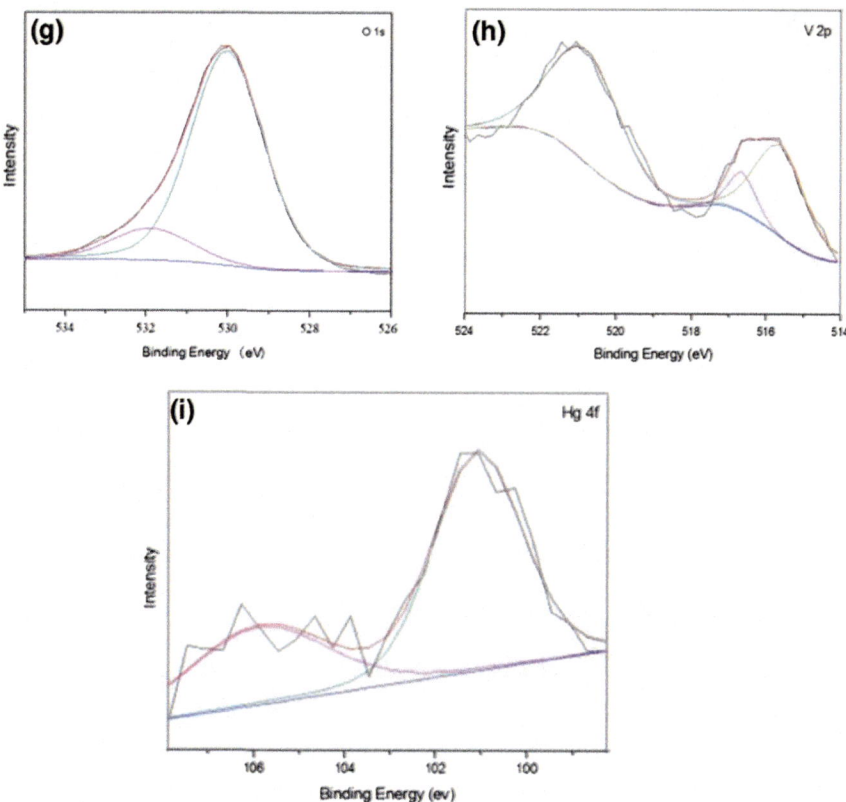

Fig. 6.6 (continued)

When the homo-hetero junctions photocatalyst system are built up, three forms of homo-hetero junctions can be formed. In order to accurately describe the enhanced photocatalytic performance, three theories are proposed. The theory I (Fig. 6.6a), the CB of V_2O_5, rutile, and anatase decreases continuously, named waterfall mechanism. The theory II (Fig. 6.6b), V_2O_5 is in the center and the CB of V_2O_5 is higher than that of rutile and anatase, which is called convex mechanism. The theory III (Fig. 6.6c), anatase is in the middle and the CB of anatase higher than that of rutile and anatase, which is called concave mechanism.

In the theory I, in the mechanism of the waterfall, these excited electrons from the CB of V_2O_5 move to rutile and eventually to anatase, while holes from VB of anatase to rutile finally reach V_2O_5. The separated electrons and holes are respectively assembled into CB of anatase and VB of V_2O_5. In the theory II, in the mechanism of the convex, these excited electrons from the CB of V_2O_5 transfer to rutile and anatase, while holes from VB of anatase and rutile arrive at V_2O_5 respectively. The separated electrons are assembled in the CB of anatase and rutile, while the separated holes are assembled in the VB of V_2O_5 respectively. In the theory III, concave mechanism, these excited electrons from the CB of rutile and V_2O_5 transfer to anatase, while holes from VB of anatase arrive at rutile and V_2O_5 respectively. The separated electrons are assembled in the CB of anatase, while the separated holes are assembled in each of rutile VB and V_2O_5. Thus, based on the proposed theories, the excited electron-hole pairs can be effectively separated in the ternary nanocomposites system. Accordingly, the energy bands adapted to homo-hetero junctions in the V_2O_5/rutile-anatase can suppress and extend the photogenerated electron-hole pairs recombination, which can lead to improved photocurrent performance and excellent photocatalytic activity. Meanwhile, the CB potential of this system is negative than the potential of $O_2/\cdot O_2^-$ couple (−0.13 eV). This suggests that the excitedly gathered electrons onto the CB of this system can be trapped by adsorbent O_2 to produce $\cdot O_2^-$ reactive species. The photogenerated holes assembled with this type of VB with strong oxidative capacities can react with adsorbed H_2O and OH^- on the surface of photocatalyst to generate $\cdot OH$. As shown in Fig. 6.6f, the CB of this system becomes negative with the potential of $O_2/\cdot O_2^-$ (−0.13 eV) pair as the doping V_2O_5 increases, indicating that the excited electrons of this system are prevented transferring to O_2. On the other hand, the VB of this system becomes negative with respect to $\cdot OH/H_2O$ (+2.27 eV) couple, and the oxidizing ability of the system weakens. When the VB of the system is under that of $\cdot OH/OH^-$ (+1.99 eV), only the photoexcited holes can oxidize the Hg^0, which is inefficient.

As shown in Fig. 6.6g, the content of Ti–O bonds is 82.1 and 86% before and after the reaction, and the content of surface-adsorbed hydroxide is 17.9 and 14% respectively. These observations mean that surface-adsorbed hydroxide species are consumed during photocatalytic oxidation process. The hydroxyl free radicals generated by surface-adsorbed hydroxide can be combined with Hg^0 to form HgO over the surface of photocatalyst. Furthermore, the increase of Ti–O bonds is due to the increase of V^{4+} after testing, which confirms that the redox couples (V^{5+}/V^{4+}) are existed on on the photocatalyst. Figure 6.6h demonstrates the V 2p XPS spectra of used V-2, which shows that compared with the fresh V-2, the binding energy of V^{5+} $2p_{3/2}$ species appears. This fact confirms that the edox couples (V^{5+}/V^{4+}) are present on the surface of as-prepared V_2O_5/rutile-anatase photocatalysts, which is present on the mercury oxidation process. In addition, the ratio of V^{5+}/V^{4+} increases after the test. It suggests that the process of V^{4+} being oxidized to V^{5+} is slower than the process of V^{5+} being converted into V^{4+}. Besides, it has been reported that V_2O_5 plays an important role in Hg^0 conversion [11]. In order to find the final product of photocatalytic reaction over the as-prepared V_2O_5/rutile-anatase nanocomposites, the Hg 4f spectra of the used V-2 is implemented and the result is shown in Fig. 6.6i, in which we can observe two peaks. The peaks at 101.09 and 105.89 eV indexed to the characteristic peaks of Hg $4f_{7/2}$ and Hg $4f_{5/2}$ for HgO respectively [12, 13]. This indicates that the main product formed on the surface of V-2 is HgO. Based on analytical and experimental results, the possible photocatalytic reaction of the V_2O_5/rutile anatase nanocomposite is given by the Eq. (6.11)–(6.27) [6, 14].

$$V_2O_5 \frac{ruile}{anatase} + hv \rightarrow V_2O_5 \frac{ruile}{anatase}(h^+ + e^-) \quad (6.11)$$

$$H_2O \leftrightarrow H^+ + OH^- \tag{6.12}$$

$$V_2O_5(h^+) + OH^-(ad) \rightarrow \cdot OH(ad) \tag{6.13}$$

$$V_2O_5(h^+) + H_2O(ad) \rightarrow \cdot OH(ad) + H^+ \tag{6.14}$$

$$O_2(g) + surface \rightarrow O_2(ad) \tag{6.15}$$

$$Anatase(e^-) + O_2(ad) \rightarrow \cdot O_2^-(ad) \tag{6.16}$$

$$O_2(ad) + H^+ \rightarrow \cdot HO_2(ad) \tag{6.17}$$

$$Anatase(e^-) + \cdot HO_2(ad) + H^+ \rightarrow H_2O_2(ad) \tag{6.18}$$

$$H_2O_2(ad) + h_v \rightarrow 2 \cdot OH(ad) \tag{6.19}$$

$$Anatase(2e^-) + \cdot HO_2(ad) + H^+ \rightarrow \cdot OH(ad) + OH^-(ad) \tag{6.20}$$

$$Hg^0(g) + surface \rightarrow Hg^0(ad) \tag{6.21}$$

$$2Hg^0(ad) + \cdot O_2^-(ad) \rightarrow 2HgO(ad) \tag{6.22}$$

$$Hg^0(ad) + \cdot OH(ad) + H^+ \rightarrow Hg^+ + H_2O \tag{6.23}$$

$$Hg^+(ad) + \cdot OH(ad) + H^+ \rightarrow Hg^{2+} + H_2O \tag{6.24}$$

$$Hg(ad) + V_2O_5 \rightarrow 2VO_2 + HgO \tag{6.25}$$

$$4VO_2 + O_2 \rightarrow 2V_2O_5 \tag{6.26}$$

$$4VO_2 + \cdot O_2^- \rightarrow 2V_2O_5 \tag{6.27}$$

6.2.2 The Photocatalytic Removal of Mercury by Carbon Spheres Supported CuO/Titanium Dioxide Photocatalysts

The influence of carbon spheres(CSs) content on the photocatalytic oxidation removal efficiency for Hg^0 was studied. Each experiment was carried out for around 180 min. First, the photocatalytic oxidation efficiency of CuO/titanium dioxide@C with different C doping ratio under ultraviolet light and LED conditions, was investigated. The results were shown in Fig. 6.7. Since the edge energy of CuO (Eg = 1.34 eV) excited by UV light was not enough to generate OH. The CuO by itself was unable to photocatalytically oxidize Hg^0 under the irradiation of ultraviolet light or xenon lamp conditions. With CuO doping, Hg removal efficiency increases up to 98%, which was much higher than that of P25 titanium dioxide under the irradiation of ultraviolet light. This was mainly because of the few Cu ions doped into the titanium dioxide lattice, improving the photocatalytic efficiency. CuO effectively suppressed electron-hole pair recombination in titanium dioxide even at moderate CuO loading [15].

With carbon spheres doping, the Hg removal efficiency of CuO/titanium dioxide@C is approximately about 79% under the UV irradiation light even low than that of CuO/titanium dioxide. But under the irradiation of LED, when C doping

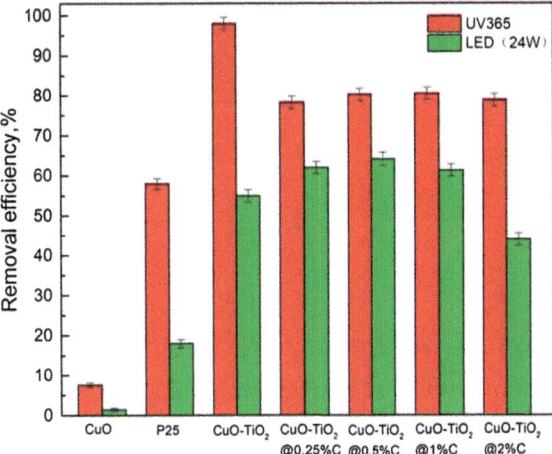

Fig. 6.7 Impact of C-doping concentration on the mercury removal efficiency of CuO/titanium dioxide@C under UV365 light and LED. Reprinted from Ref. [20], Copyright 2015, with permission from Elsevier

amount reached 0.5 wt%, the removal efficiency reached 64.1%, which was higher than that of CuO/ titanium dioxide and P25. The reason is that, at low CuO loading (0.31 wt%), copper was existed as a submonolayer Cu(II) or CuO species on titanium dioxide. The Fermi level for such adsorbed species was positive with respect to the conduction band of titanium dioxide, causing the movement of photo-excited electrons from titanium dioxide to the surface Cu(II) or CuO centers. At the same time, when CSs absorbs irradiation as a photosensitizer and inject the photogenerated electrons into the conduction band of CuO/titanium dioxide, then the rate of electron transferring to oxygen adsorbed on the catalyst surface increases. Simultaneously, the positively charged CSs might remove electrons from the valence band of CuO/titanium dioxide, making them leave holes. However, with doping the doping of CSs above 0.5 wt%, the removal efficiency of Hg^0 decreased. The reason is that increasing the loading ratio of CSs, the valence band level of CuO/titanium dioxide@C became negative with respect to OH/OH^- (+1.99 eV) couple, weakening the ability of oxidation capacity. This result showed that the best Hg^0 oxidation efficiency was observed for the CuO/titanium dioxide@0.5 wt% C samples.

Under the light excitation, titanium dioxide was active for Hg^0 photocatalytic oxidation. The main reason was that titanium dioxide generated photo-excited electrons and photoexcited holes, which could convert the water vapor adsorbed on the catalyst surface into H^+ and OH^- [16]. After that photo-excited holes converted OH^- into $\cdot OH$, which was a strong oxidant. O_2 adsorbed on the catalyst surface was reduced by the photo-excited electrons to generate superoxide radicals $\cdot O_2^-$. Hg^0 can be oxidized to Hg^{2+} by both $\cdot O_2^-$ and $\cdot OH$ [17]. The photocatalytic oxidation steps can be described as Eqs. (6.28)–(6.35) [18, 19].

$$Carbon \frac{CuO}{TiO_2} + h\upsilon \rightarrow Carbon \frac{CuO}{TiO_2}(h^+ + e^-) \quad (6.28)$$

$$OH^- + h^+ \rightarrow \cdot OH \quad (6.29)$$

$$O_2 + 2H^+ + 2e^- \to H_2O_2 \tag{6.30}$$

$$H_2O_2 + h\upsilon \to 2 \cdot OH \tag{6.31}$$

$$O_2(g) + e^- + \text{surface} \to \cdot O_2^-(ad) \tag{6.32}$$

$$Hg^0(g) + surface \to Hg^0(ad) \tag{6.33}$$

$$3Hg^0(ad) + 2 \cdot O_2^-(ad) + 2H^+ \to 3HgO(ad) + H_2O \tag{6.34}$$

$$Hg^0(ad) + 2 \cdot OH(ad) \to HgO + H_2O \tag{6.35}$$

6.2.3 The Photocatalytic Removal of Mercury by Carbon Decorated In_2O_3/Titanium Dioxide Photocatalysts

6.2.3.1 Effect of Carbon

The visible-light absorption performance greatly affects photocatalytic property of photocatalyst. UV-visible diffuse reflectance spectroscopy (DRS) study is performed to analyze the optical properties of the as-prepared nanocomposites. As can be seen in Fig. 6.8a, the absorption edges of carbon modified P25 and In/Ti-2 shift to the visible light range. In addition, carbon modified samples has a distinct enhancement for light absorption ability in the range of UV and visible light compared with P25, reaction can be describedwhich certified the potential application of the carbon modified samples as visible light active photocatalysts. The enhanced absorption ability may be attributed to the new doping energy levels formed by carbon doping [21]. The band gap energy (E_g) of the as-prepared ternary nanocomposite can be calculated according to Kubelka-Munk equation:

$$E_g = \frac{1240}{\lambda_{Absorp.Edge}} \tag{6.36}$$

In the prepared samples of P25, C-P25, In/Ti-2, and C-In/Ti-2, the E_g values were found to be 3.21, 3.06, 3.06 and 3.05 eV respectively. Compared with previous study [22], the reason of narrowed band gap (BG) can be attributed to the doped carbon acting as sensitizer, which can absorb visible light and enhance synergistic properties by joint electronic system formed between titanium dioxide and carbon. In addition, carbon doping can promote the formation of chemical bonds of O-Ti-C, introducing seized locality states into the BG of titanium dioxide lattice resulting in narrow BG narrowed BG, which enhances visible light absorption property and promotes the more efficient utilization of light for the photocatalysis [23].

In general, the photocatalytic activity of ternary nanocomposites greatly depends on the separation efficiency of the photoinduced electrons and holes. Photoluminescence (PL) emission spectra are employed to characterize the separation capability of photoexcited electrons and holes. Lower PL emission intensity means higher photogenerated electrons and holes separation capability. The separated electrons and holes

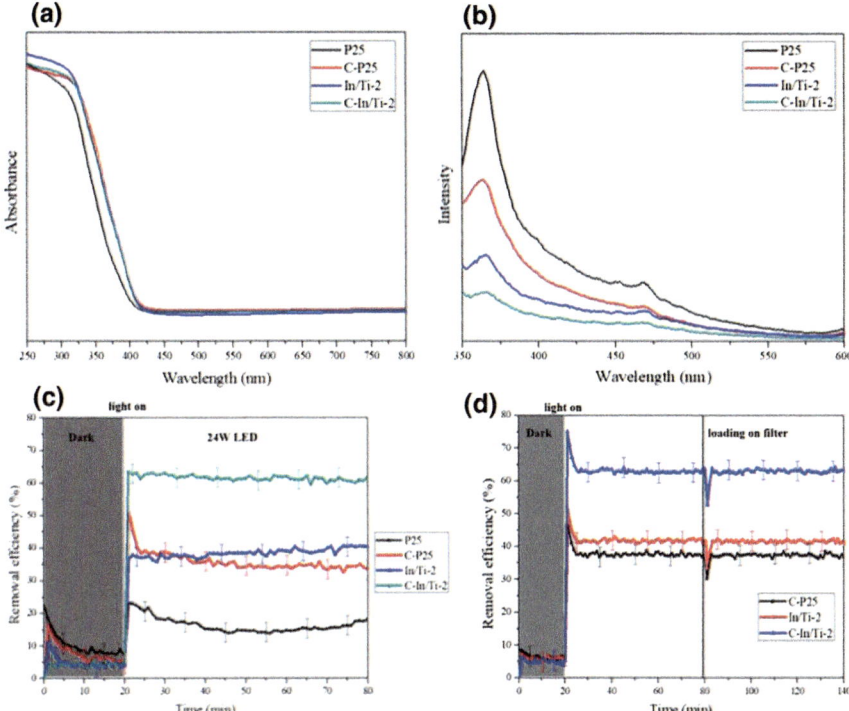

Fig. 6.8 a UV-vis diffuse reflectance spectroscopy, **b** Photoluminescence emission spectra, **c** Effects of carbon on Photocatalytic removal of Hg^0, **d** Removal efficiency before and after loading light filter ($\lambda < 350$ nm). Reprinted from Ref. [25], Copyright 2017, with permission from Elsevier

combine with hydroxyl and oxygen to produce active component improving photocatalytic performance [24]. Figure 6.8b shows PL emission spectra of as-prepared P25, C-P25, In/Ti-2, and C-In/Ti-2 samples, which demonstrates that PL emission intensity of C-P25 is lower than that of P25, indicating that the excited electrons and holes are efficiently separated with carbon doping. In addition, It also shows that doping carbon in In/Ti-2 sample lowers PL emission intensity and effectively promotes separation of photoexcited electron-hole pair.

Based on the analysis, doping titanium dioxide with carbon not only can narrow band gap but also inhibit recombination of the photogenerated charge carriers. Thus the as-prepared samples with carbon doping can behave higher photocatalytic activity than pure samples without carbon. Hg^0 is removed using the prepared samples (P25, C-P25, In/Ti-2, and C-In/Ti-2) in order to detect photocatalytic properties. The photocatalytic properties of ternary nanocomposites were investigated under visible light (24 W LED). Each test was conducted for about 80 min including under dark (20 min) and visible light irradiation (60 min) respectively. The photocatalytic efficiencies are shown in Fig. 6.8c, which indicates that the removal efficiency of C-P25 is higher than that of P25, and the efficiencies are 17.5 and 37.2% respectively. On the other hand, the removal efficiency of C-In/Ti-2 is higher than that of In/Ti-2, and efficiencies of In/Ti-2 and C-In/Ti-2 are 39.6 and 61.1% respectively. Therefore, by doping carbon, the photocatalytic performance of the sample can be enhanced. The reason may be that doping carbon can narrow band gap, confirmed by UV-vis DRS analysis, and promote the separation of charge carriers, which is verified by PL analysis. Meanwhile, the experiment with a light filter ($\lambda < 350$ nm) loaded on the LED was carried out. The experimental results are shown as Fig. 6.8d, and the removal efficiency was not changed after loading with the light filter ($\lambda < 350$ nm).

6.2.3.2 Effect of In_2O_3 Content

The UV-vis DRS of ternary nanocomposites are shown in Fig. 6.9a. Because of the large E_g (3.20 eV for anatase, 3.02 eV for rutile), titanium dioxide only absorbs UV light. Figure 6.9a indicates that the absorption edges of C-In/Ti-X (X = 1, 2, 3, 6, and 9) nanocomposites shift to the visible light range. On the other hand, the absorption edges of as-prepared nanocomposites gradually shift to lower energy with the added amount of In_2O_3 loading. According to Kubelka-Munk equation, the BG of C-In/Ti-X (X = 1, 2, 3, 6, and 9) are calculated. The E_g values are found to be 3.21, 3.05, 3.05, 3.05, 3.04, and 3.03 eV for P25, C-In/Ti-1, C-In/Ti-2, C-In/Ti-3, C-In/Ti-6, and C-In/Ti-9 respectively. The calculation results show that the absorption capacity improves with the increasing of the content of In_2O_3. It is clear that In_2O_3 is successfully embedded in the titanium dioxide host nanoparticles to form a heterojunction, which changes its electronic and optical properties, and promotes the separation and migration of photogenerated electrons-holes at the interface. Meanwhile, carbon modification and In_2O_3 doping can increase absorbance of as-prepared photocatalysts to visible region. Thus, this is a promising method to synergize with In_2O_3 and carbon to increase the photoactivity of the support and broaden absorbance within the visible region.

Photoluminescence (PL) emission spectra of C-In/Ti-X (X = 1, 2, 3, 6, and 9) are shown in Fig. 6.9b. It shows that PL intensity of the samples C-In/Ti-X (X = 1, 2, 3, 6, and 9) are lower than that of P25, which shows that the recombination of photoexcited electrons and holes are suppressed because of the formed heterojunctions between In_2O_3 and titanium dioxide, promoting migration of photogenerated electrons and holes. In addition, C-In/Ti-2 sample has the lowest PL intensity of the as-prepared samples, which not only shows that the proper doping level of In_2O_3 and carbon can provide appropriate capture traps but also suggests that the efficient separation and migration of electrons and holes are promoted because of the cooperative effect of carbon modification and heterojunctions between In_2O_3 and titanium dioxide. The separated holes combining with OH^- produce ·OH radicals, which increase the photocatalytic activity [26]. However, at high doping levels, extra In_2O_3 accumulates on the surface of titanium dioxide, and acts as recombination centers increasing the recombination of photo-production carriers, and in turn lowering photocatalytic activity.

6.2 The Photocatalytic Removal of Mercury by Metal or Nonmetal … 119

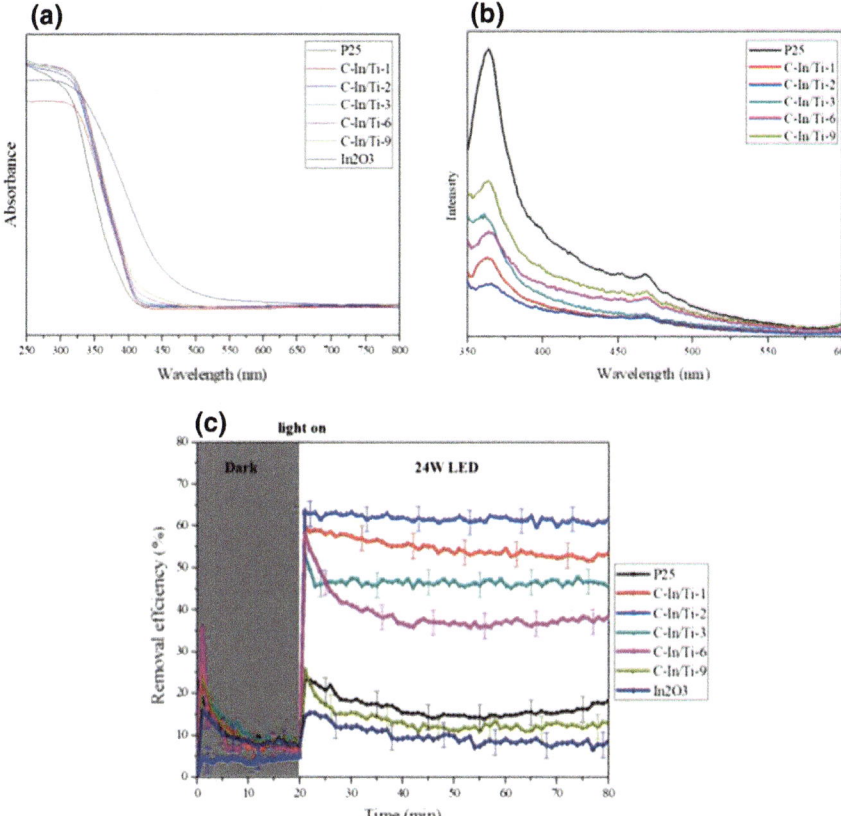

Fig. 6.9 **a** UV-vis diffuse reflectance spectroscopy, **b** Photoluminescence emission spectra, **c** Effects of In_2O_3 on photocatalytic removal of Hg^0. Reprinted from Ref. [25], Copyright 2017, with permission from Elsevier

Based on the analysis, carbon modified In_2O_3/titanium dioxide heterostructure not only can narrow band gap but also enhance the charge transfer and separation. Therefore the as-prepared carbon modified In_2O_3/titanium dioxide heterostructure can exhibit improved photocatalytic performance compared to pure P25. The photocatalytic activities of as-prepared samples C-In/Ti-X (X = 1, 2, 3, 6, and 9) were conducted under visible light (24 W LED). The results are shown in Fig. 6.9c, which shows that the Hg^0 removal efficiency over as-prepared samples C-In/Ti-X (X = 1, 2, 3, 6, and 9) increases first and then decreases with the content of In_2O_3 increasing, and that the bending point is 2 mol%. The optimal molar ratio of In/Ti is 2 mol%, and the highest photocatalytic efficiency is about 61.1%, which is much higher than that of P25 under visible light irradiation. Therefore, synergizing In_2O_3 and carbon doping into titanium dioxide is an efficient method to improve photoactivity. The enhanced activity is due to that doping titanium dioxide with carbon can form oxygen vacancies and introduce an additional carbon-doping level upon the VB of titanium dioxide support [27]. Further, by doping In_2O_3 in titanium dioxide, a heterojunction is formed between titanium dioxide and In_2O_3 and a surface state energy level is introduced, which can reduce the BG. Under visible light irradiation, photoexcited electrons move to the conduction band (CB) of titanium dioxide, meanwhile, photoexcited holes transfer to the valence band (VB) of In_2O_3, which commence at titanium dioxide. The process can efficiently hinder the recombination between photoproduced electrons and holes and aliment electron-hole separation and migration at the interfaces. Doping titanium dioxide with carbon can increase S_{BET} of host. Co-doping titanium dioxide with In_2O_3 and carbon, it is possible to increase the visible light absorption characteristics and prolong the lifetime of photoexcited charge carriers. The BET surface area, ability to separate photoexcited electrons and holes and absorption boundary all affect the photocatalytic reaction. The high mercury removal efficiency is attributed to the combined action. The sample C-In/Ti-2 has the highest BET surface area, the photo-excited electron and hole separation ability and the absorption boundary is stronger so that C-In/Ti-2 has the highest Hg removal efficiency.

6.2.3.3 The Stability of Photocatalyst

The cyclic stability is an important factor for practical application of photocatalyst. The cyclically photocatalytic oxidation of Hg^0 was performed to detect the photocatalytic stability to verify the cyclically stable ability of the as-prepared ternary nanocomposite. Five cyclic tests were conducted with C-In/Ti-2 and each cyclic test was performed under the uniform conditions shown in Fig. 6.10a. In the test, the quartz glass loaded with C-In/Ti-2 photocatalyst (0.08 g) was first detected under dark for 20 min and visible light was turned on below. After irradiating for 30 min under visible light, the LED lamp was turned off and the followed cyclic experiment was conducted after 10 min. The same procedures were carried out in the followed several experiments. The photocatalytic activity of C-In/Ti-2 remained almost unchanged even after five consecutive cycle experiments and showed high stability. As can be seen from Fig. 6.10b, the reduced amplitude of removal efficiency after light was turned off is consistent with the decreased amplitude of temperature (room temperature is 29.8 °C). Therefore, the residual reactivity might be related to temperature. When light was turned on, photocatalysis and thermal catalysis are performed at the same time. Since the temperature of the photocatalytic reactor is low (the temperature is 61 °C after three hours irradiation), thermal catalysis efficiency is very small. When the light was turned off, only thermal catalysis is performed, and the efficiency decline slows down. This residual reactivity lasts about 40 min before returning to an initial state. The structures of the C-In/Ti-2 sample before and after five consecutive cycle experiments are also characterized by the XPS analysis, as exhibited in Fig. 6.10c. XPS patterns of the C-In/Ti-2 sample before and after photocatalytic oxidation of Hg^0 show no observable change.

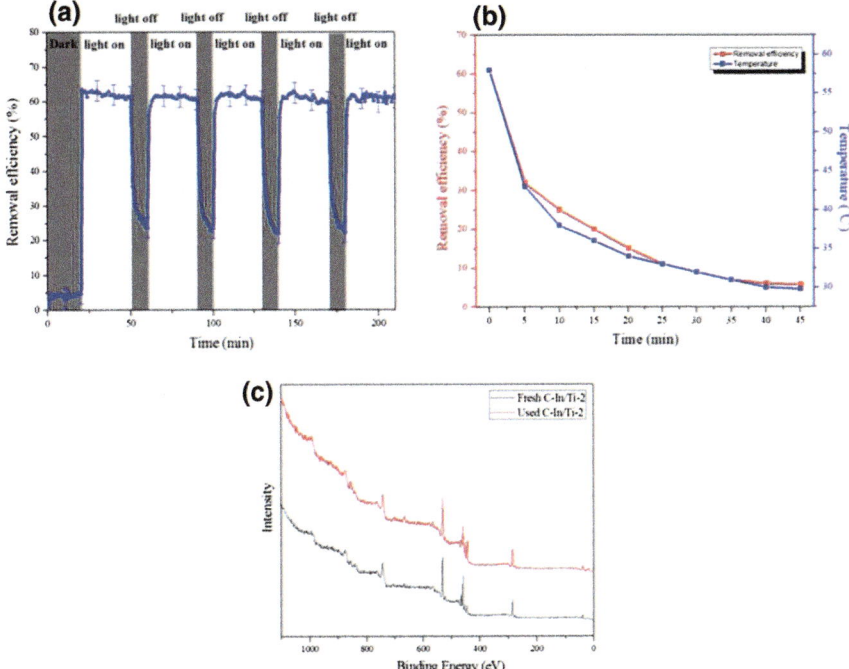

Fig. 6.10 a Photocatalytic activity of the C-In/Ti-2 sample with five times of cycling experiments, **b** the relationship between removal efficiency and temperature, **c** XPS survey spectra of the C-In/Ti-2 sample before and after the photocatalytic experiments. Reprinted from Ref. [25], Copyright 2017, with permission from Elsevier

6.2.3.4 The Possible Charge Transfer Mechanism (CTM) in Carbon Modified In_2O_3/Titanium Dioxide Heterostructure

Based on the experimental and characterization results, we propose a possible charge transfer mechanism (CTM) to reveal the enhanced photoactivity of the carbon modified In_2O_3/titanium dioxide heterostructure. As shown in Fig. 6.11a, because of its large energy gap (3.20 eV for anatase, 3.02 eV for rutile), titanium dioxide can only absorb UV light. However, when titanium dioxide is doped with carbon additional energy levels above the valence band, its absorption boundary will be expanded to visible light. Under light irradiation, electrons are inspired from the VB of titanium dioxide and migrate to (101) plane due to the formed surface heterojunction between (101) and (001) faces of titanium dioxide shown in Fig. 6.11b [27]. Meanwhile, the CB bottom of titanium dioxide (In_2O_3) lies at −0.18 eV (−0.63 eV), and the valence band (VB) top of titanium dioxide (In_2O_3) lies at 3.02 eV (2.17 eV) [28], which indicate that CB bottom and VB top of titanium dioxide are lower than that of In_2O_3 respectively, shown as Fig. 6.11c. Thus, the coupled ternary nanocomposites form an efficient heterojunction, which can be triggered under visible light

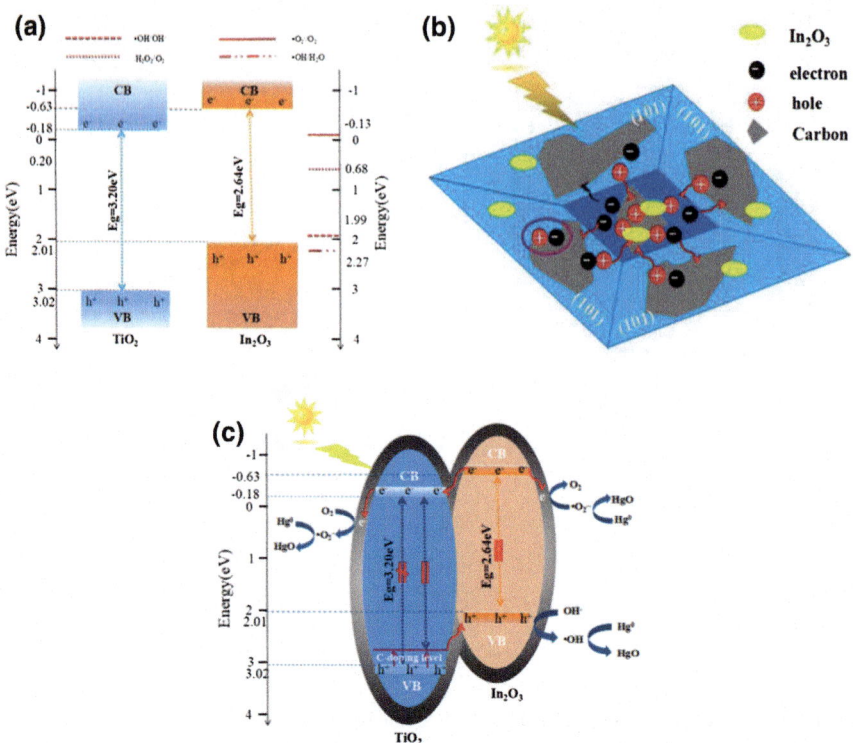

Fig. 6.11 a The CB and VB levels of titanium dioxide and In_2O_3, **b** Schematic diagram of the possible charge transfer mechanism between different plane, **c** Schematic diagram of the possible charge transfer mechanism of excited electrons and holes. Reprinted from Ref. [25], Copyright 2017, with permission from Elsevier

irradiation. The electrons are inspired from the VB of In_2O_3 and titanium dioxide, generating corresponding number of electropositive holes at the VB. The photoexcited electrons on the CB of In_2O_3 can easily transfer to the CB of titanium dioxide and then transfer to the modified carbon. At the same time, carbon covering the surface of In_2O_3/titanium dioxide heterostructure can incorporate the electrons on the CB of In_2O_3 [27]. Similarly, the holes on the VB of titanium dioxide can easily move to the VB of In_2O_3. On the other hand, because of highly dispersed In_2O_3 on the titanium dioxide hosts, the Schottky barrier is formed at the interface between titanium dioxide and In_2O_3, which can hinder the recombination of electron and hole pairs. As a result, the excited charges can be efficiently separated and migrate in carbon modified In_2O_3/titanium dioxide heterostructure to promote the movement of excited charges at the interfacial face, which is confirmed by PL emission spectra, causing higher photocatalytic activity.

6.2.3.5 The Possible Photocatalytic Reaction Mechanism (PCRM) of Carbon Modified In$_2$O$_3$/Titanium Dioxide Heterostructure

To explore the possible photocatalytic reaction mechanism (PCRM) of carbon modified In$_2$O$_3$/titanium dioxide heterostructure, the C-In/Ti-2 sample is investigated by XPS analysis. As can be seen in Fig. 6.12a, the O 1s peak at 529.5 eV can be corresponded to titanium dioxide lattice oxygen, the peak at 530.7 eV can be indexed to oxygen anions from In$_2$O$_3$, the peak at 531.6 eV can be corresponded to oxygen bound species C-O, and the peak at 532.5 eV is the characteristics of the surface hydroxyl groups. The lattice oxygen contents before reaction and after reaction are 63.8 and 33.0% respectively, while the contents of oxygen bound species C–O are 7.2 and 27.2% respectively. It indicates that lattice oxygen is consumed during the photocatalytic reaction (PCR). The phenomenon of increased oxygen bound species C-O is verified by high resolution C 1s spectrum of the used C-In/Ti-2 sample, shown in Fig. 6.12b. In addition, the high resolution spectrum of Ti 2 p for the C-In/Ti-2 samples used is also investigated. As shown in Fig. 6.12c, two peaks assigned to Ti^{3+}, which are not found for fresh C-In/Ti-2 sample, indicate that Ti^{4+} varies to Ti^{3+} during the PCR. Furthermore, high resolution spectra of Hg 4f shown in Fig. 6.12d demonstrate two peaks at 101.7 and 103.3 eV corresponding to Hg 4f$_{7/2}$ and Hg 4f$_{5/2}$ of HgO respectively, evidencing that the product of PCR is HgO [28, 29]. On the other hand, XRD patterns of the C-In/Ti-2 sample before and after photocatalytic oxidation of Hg^0 (Fig. 6.12e) indicate that the diffraction peak centered at $2\theta = 25.3°$ corresponding with anatase (101) plane shows slight weakness after this reaction, while other peaks have no changes. This indicates that anatase (101) plane has several changes during PCR.

Considering the experimental results and XPS and XRD analysis, the possible PCRM is shown in Fig. 6.12f. First, the energy of visible light exceeds the BG of In$_2$O$_3$ and carbon-doped titanium dioxide, so visible light irradiation can trigger the formation of electron-hole pairs. Thereafter, according to CTM, photogenerated electron-hole can be efficiently separated and photogenerated electrons mainly gathered on anatase (101) plane. Abundantly separated photogenerated electrons reduce O$_2$ adsorbed on the surface of ternary nanocomposites to produce active component ·O$_2^-$ (the potential of O$_2$/ · O$_2^-$ is −0.13 eV), which can oxidize Hg into HgO. At the same time, the reactive holes generated at the VB of this ternary material has strong oxidative capacities and can react with adsorbed H$_2$O and OH$^-$ to produce ·OH (the potential of ·OH/H$_2$O and ·OH/OH$^-$ are +2.27 and +1.99 eV), which can oxidize Hg also. Because of its higher oxidization ability than that of ·O$_2^-$, ·OH are consumed more than ·O$_2^-$ during PCR, which results in abundant electrons gathering at anatase (101) plane [30, 31]. Ti^{4+} is reverted to Ti^{3+} by the gathered electrons and releases lattice oxygen (O^{2-}) to oxidize Hg into HgO, and O^{2-} decreases during its oxidizing mercury process, which is evidenced by O 1s high spectrum of the used C-In/Ti-2 sample. Meanwhile, Ti^{3+} is oxidized into Ti^{4+} by ·O$_2^-$, but oxidation process is less than reduction process, leading to the amount of Ti^{3+} increased, which is confirmed by Ti 2p high spectrum of the used C-In/Ti-2 sample shown in Fig. 6.12g. As shown in Fig. 6.12h, the main active component (·OH) can be produced in two

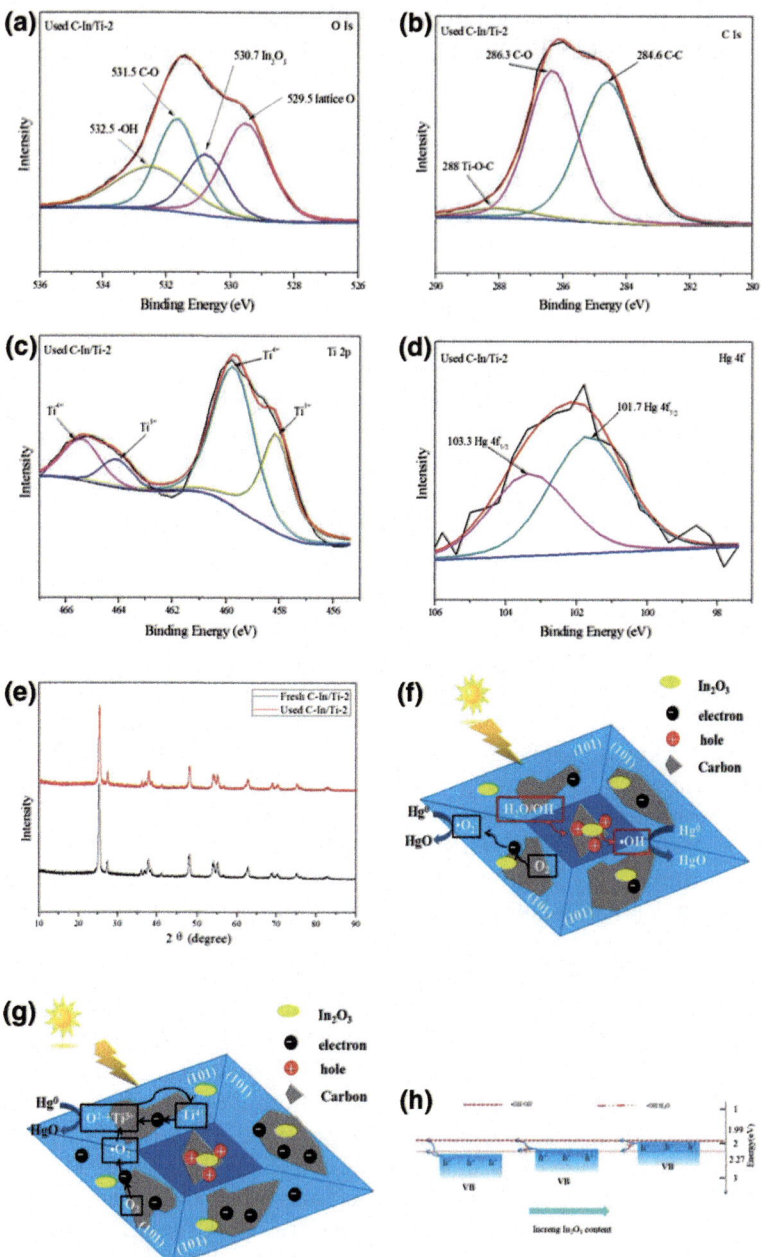

Fig. 6.12 High resolution XPS spectra of used C-In/Ti-2 for **a** O 1s, **b** C 1s, **c** Ti 2p, **d** Hg 4f, **e** XRD patterns of fresh and used C-In/Ti-2 samples, **f** Schematic diagram of the possibly photocatalytic mechanism of the carbon modified In_2O_3/titanium dioxide heterostructure, **g** Schematic diagram of the possible anatase (101) plane Ti valence transformation, **h** Impact of In_2O_3 on the VB levels of the ternary nanocomposites. Reprinted from Ref. [25], Copyright 2017, with permission from Elsevier

6.2 The Photocatalytic Removal of Mercury by Metal or Nonmetal …

ways, i.e., water and hydroxyl are oxidized to ·OH (the potential of ·OH/H_2O and ·OH/OH^- are +2.27 and +1.99 eV respectively). When the ternary material generated, the potential of CB for the system is lower than potential of titanium dioxide and higher than potential of In_2O_3. As the content of In_2O_3 increases, the potential of CB relative to the system keeps decreasing. When the In/Ti molar ratios are below 0.02:1, the potential of CB is slightly higher than that of ·OH/H_2O couple. As the In/Ti molar ratios increase, the potential of CB is lower than that of ·OH/H_2O couple, leading the ·OH production decreases. As the active component decreases, photoexcited holes and electrons recombination and S_{BET} decreases, so that the photocatalytic performance is deteriorated. In summary, possible PCRM involves two main reaction process: producing process of active components [Eqs. (6.37)–(6.41)] and photocatalytic oxidation process [Eqs. (6.42)–(6.46)].

$$Carbon\frac{In_2O_3}{TiO_2} + h\upsilon \rightarrow Carbon\frac{In_2O_3}{TiO_2}(h^+ + e^-) \quad (6.37)$$

$$OH^- + h^+ \rightarrow \cdot OH \quad (6.38)$$

$$O_2 + 2H^+ + 2e^- \rightarrow H_2O_2 \quad (6.39)$$

$$H_2O_2 + h\upsilon \rightarrow 2 \cdot OH \quad (6.40)$$

$$O_2(g) + e^- + surface \rightarrow \cdot O_2^-(ad) \quad (6.41)$$

$$Hg^0(g) + surface \rightarrow Hg^0(ad) \quad (6.42)$$

$$3Hg^0(ad) + 2 \cdot O_2^-(ad) + 2H^+ \rightarrow 3HgO(ad) + H_2O \quad (6.43)$$

$$Hg^0(ad) + 2 \cdot OH(ad) \rightarrow HgO + H_2O \quad (6.44)$$

$$Hg(ad) + 2TiO_2 \rightarrow Ti_2O_3 + HgO(ad) \quad (6.45)$$

$$3Ti_2O_3 + 2 \cdot O_2^- + 2H^+ \rightarrow 6TiO_2 + H_2O \quad (6.46)$$

6.3 The Photocatalytic Removal of Mercury by Other Photocatalysts

6.3.1 The Photocatalytic Removal of Mercury by $BiOIO_3$

6.3.1.1 Hg^0 Removal Under LED and UV Light Irradiation

The photocatalytic activities of the prepared sample were roughly calculated by photocatalytic oxidation of Hg^0. Using a 24-watt LED illumination as a light source, its wavelength was 420 nm, which is exactly the range of visible light. Each experiment was performed under dark conditions for 1000 s, 3500 s under visible light, and 2700 s under ultraviolet light for a total of 7200 to study the photocatalytic oxidation of Hg^0 (Fig. 6.13). Under the visible light of LED lamp, the removal rate of Hg^0 in the original $BiOIO_3$ and BiOI was only 59.31 and 12.35% respectively,

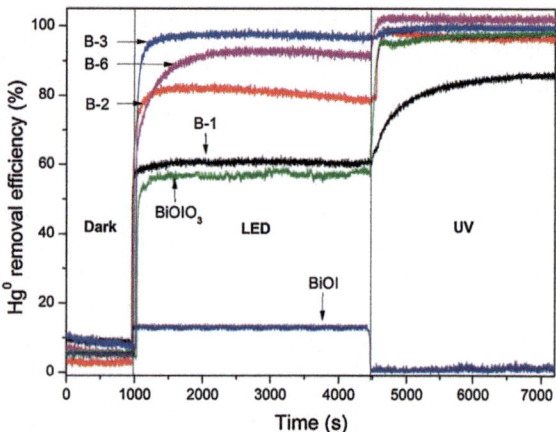

Fig. 6.13 Removal efficiencies of Hg^0 for the samples under dark, LED and UV light irradiation. Reprinted from Ref. [32], Copyright 2017, with permission from Elsevier

but the $BiOI/BiOIO_3$ heterostructures had very good photocatalytic activity. The removal rate of Hg^0 at first increased with the increasing content of BiOI, and then it decreased as the BiOI content continually increased. It can be found that when the molar ratio of $BiOI/BiOIO_3$ compound was 3:1 (B-3), the peak removal efficiency of Hg^0 was about 98.53%, which was in full agreement with the characterization results. However, the photocatalytic activity of B-3 was not the highest under ultraviolet irradiation. On the contrary, the photocatalytic oxidation efficiency of B-6 was very high, even up to 100%. Due to the instability of BiOI, BiOI hardly showed any photocatalytic activity. BiOI was composed of alternating layers of $[Bi_2O_2]$ flakes and double I flakes, which could be stacked together through non-bonding interactions [32], but their interaction was relatively weak, so this layered structure was compared. The separation phenomenon easily occurs along the (001) direction.

6.3.1.2 Stability of BiOI/BiOIO$_3$ Heterostructures

Besides photocatalytic activity, photocatalytic stability is also very significant. Under the illumination of LED light, each group was subjected to a Hg^0-removal cycle test with a duration of approximately 2500 s (Fig. 6.14a). This was to study whether B-3 photocatalyst was stable and studied for 7 consecutive cycles. Hg^0 removal efficiency of B-3 was studied. In each cycle, first the quartz glass coated with B-3 photocatalyst was exposed to about 2500 s under the LED light, then the LED was turned off and then the next cycle was performed when the Hg^0 concentration reached a certain value and tended to be stable. Experiments were conducted in strict accordance with the above steps. After 7 consecutive cycles, the removal rate of Hg^0 could be as high as 98.5%, indicating that B-3 is an excellent photocatalytic activity catalyst capable of effectively removing Hg^0. As shown in Fig. 6.14b, the main (001) peak of the B-3 photocatalyst was maintained at a fixed value before and after 7 consecutive cycles,

6.3 The Photocatalytic Removal of Mercury by Other Photocatalysts

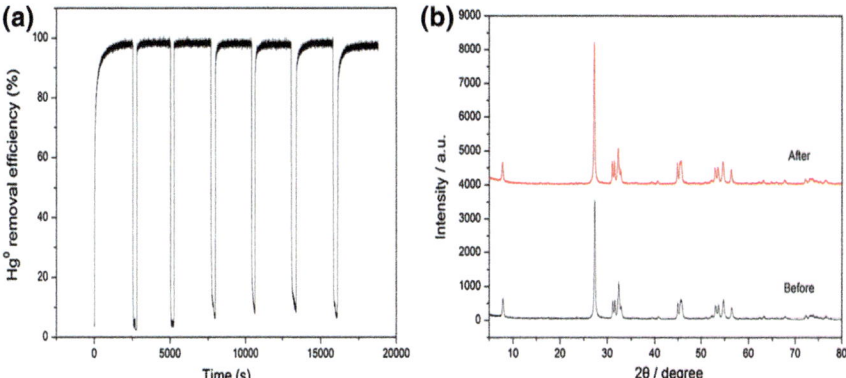

Fig. 6.14 Repeated photocatalytic activity of B-3 (**a**) and the XRD comparison spectrum (**b**). Reprinted from Ref. [32], Copyright 2017, with permission from Elsevier

indicating that the B-3 photocatalyst has good stability, even though after 7 cycle experiments it also had stable chemical structure.

6.3.1.3 Photocatalytic Reaction Mechanisms

The photocatalytic activities of the prepared BiOI/BiOIO$_3$ heterostructures are higher than that of any single component, and this heterostructure can make the separation of e$^-$ and h$^+$ more efficient. The energy band structure of the semiconductor and the direction of e$^-$ and h$^+$ transport between the semiconductors are closely related. The Fermi level of p-type semiconductor BiOI is close to the valance band, however the Fermi level of n-type semiconductor BiOIO$_3$ is approach to the conduction band. When a p-n junction is formed between the BiOIO$_3$ and BiOI, the Fermi level of p-type semiconductor BiOI and n-type semiconductor BiOIO$_3$ will reach an equilibrium in the beginning, meanwhile the valance and conduction band positions of BiOI move towards the more negative electronegativity, and the valance and conductor bands position of BiOIO$_3$ move towards the more positive electronegativity. It can be seen from Fig. 6.15a that BiOI excited e$^-$ and h$^+$ under the irradiation of visible light, while the conduction position of BiOI is more negative than that of BiOIO$_3$, so the e$^-$ moves from the conduction band of BiOI to the conduction band of BiOIO$_3$, however the h$^+$ remains in the valance band of BiOI. The energy of e$^-$ is higher. After they are transferred to the conduction band of BiOIO$_3$, they can react with oxygen adsorbed on the surface of the catalyst to generate ·O$_2^-$O$_2^-$. The ·OH can be formed by the h$^+$ left on the valance band of BiOI through oxidizing the OH$^-$ or H$_2$O adsorbed on the surface of catalyst, and then the Hg^0 can be oxidized by high activity ·OH and ·O$_2^-$ photocatalytically into Hg^{2+}. What's more, because of the internal polar field in the original BiOIO$_3$ and BiOI, the e$^-$ and h$^+$ migrate to the composite interface of the {010} facets of BiOIO$_3$ and the {001} facets of

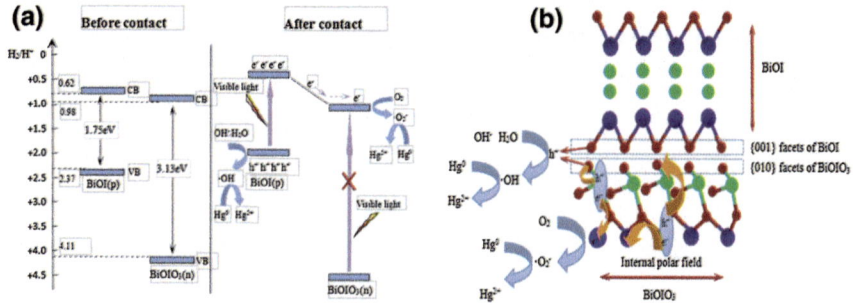

Fig. 6.15 **a** Mechanism of separation of photo electron-holes. **b** Schematic diagram of composite interface structure of BiOI/BiOIO₃ composites. Reprinted from Ref. [32], Copyright 2017, with permission from Elsevier

BiOI. There is an inside polar field perpendicular to the nano sheet in BiOI and an inside polar field parallel to the nano sheet in BiOIO₃ [33]. As can be seen from Fig. 6.15b, due to the vertical relationship between the internal polar field and the nanosheet, the h⁺ in the BiOI migrates from the interior to the {001} plane, while the h⁺ photogenerated at IO₃ pyramid and BiO₆ pyramid in BiOIO₃ tend to spread around the {010} facets due to the horizontal inside polar field and the e⁻ generated at IO₃ pyramid and BiO₆ pyramid transfer to Bi 6p CB bottom. In this process, the horizontal transition of h⁺ on the {010} plane of BiOIO₃ is relatively short relative to the other directions, and therefore it is easier to contact the material in the horizontal direction. As being reported on most of literatures [34–36], both the h⁺ and ·O₂⁻ are the key species of the oxidation. Because the ·OH oxidizes Hg⁰ into Hg²⁺, and the ·OH is formed with the h⁺ reacting with H₂O or OH⁻, so the indispensable species of reacting oxidation is still h⁺.

The Photocatalytic oxidation reaction can be described as the following Eqs. (6.47)–(6.57) [37–39].

$$BiOI/BiOIO_3 + h\upsilon \rightarrow BiOI/BiOIO_3 + h^+ + e^- \tag{6.47}$$

$$H_2O \leftrightarrow H^+ + OH^- \tag{6.48}$$

$$OH^-_{ad} + h^+ \rightarrow \cdot OH_{ad} \tag{6.49}$$

$$H_2O_{ad} + h^+ \rightarrow \cdot OH_{ad} + H^+ \tag{6.50}$$

$$O_{2ad} + e^- \rightarrow \cdot O^-_{2\,ad} \tag{6.51}$$

$$O_{2ad} + H^+ \rightarrow \cdot HO_{2ad} \tag{6.52}$$

$$\cdot HO_{2ad} + e^- + H^+ \rightarrow H_2O_{2ad} \tag{6.53}$$

$$H_2O_{2ad} + h_\upsilon \rightarrow 2 \cdot OH_{ad} \tag{6.54}$$

$$2Hg^0_{ad} + \cdot O^-_{2\,ad} \rightarrow 2HgO_{ad} \tag{6.55}$$

$$Hg^0_{ad} + \cdot OH_{ad} + H^+ \rightarrow Hg^+ + H_2O \tag{6.56}$$

$$Hg_{ad}^+ + \cdot OH_{ad} + H^+ \rightarrow Hg^{2+} + H_2O \tag{6.57}$$

6.3.2 The Photocatalytic Removal of Mercury by CSs-BiOI/BiOIO$_3$

6.3.2.1 Photocatalytic Oxidation of Gaseous Hg^0

In order to investigate the Hg^0 removal performance of pure BiOIO$_3$ and CSs-BiOI/BiOIO$_3$, they were performed under dark conditions for 10 min and under light conditions for 60 min(Fig. 6.17). In order to produce a comparative experiment for convenient testing, a blank test without photocatalysts was performed. As shown in Fig. 6.17a, blank experiment of about 90 min was performed under LED illumination and the average Hg^0 concentration was found to be 0.5% lower than that in dark conditions. Under dark for 10 min before illumination, it was to attain the adsorption-desorption equilibrium between the Hg and the photocatalysts. The adsorption efficiency of all samples is listed in detail in Table 6.1 and it can be seen that the efficiency increases significantly with the increasing of BET surface area. Figure 6.17b shows the Hg removal efficiency of different photocatalysts under two kinds of illumination, and the concrete results are shown in Table 6.1. For the P-BiOIO$_3$, η_1 of 73.7% is observed, showing that the pure BiOIO$_3$ can remove gas Hg^0 very efficiently under the irradiation of LED light. Figure 6.16a shows that the LED light around 410 nm can irritate free electrons from valence band of pure BiOIO$_3$ very weakly. Although only a few of electrons can be stimulated, the heterolayered structure and inside polar field causes efficient separation of photogenerated electron-hole pairs and the oxidation ability of positive holes is very active, bringing about good efficiency of Hg^0 removal. The mercury removal efficiency increases first and then decreases with the increase of CSs doping amount. The as-prepared CSs-BiOI/BiOIO$_3$ possessed localized occupied states in the gap, extending photoresponse into the visible range above 420 nm. Therefore, the CSs-BOI-1 under the illumination of the LED can be effectively excited, resulting in relatively large energy, and the removal rate of Hg^0 is also very high, which can reach 92.9%. Interestingly, we can see that the Hg removal efficiency by CSs-BiOI/BiOIO$_3$ with weight ratios of CSs more than 3% is higher than that by CSs-BOI-3 (Fig. 6.17b), and it can be attributed that appropriate BiOI improves the segmentation of electron-hole pairs and their BET surface areas increase. It is known that a catalyst with a larger surface area will provide more surface active sites for the adsorption of the reactant molecules, which will allow it to have a better adsorption capacity and a higher photocatalytic activity [41, 42].

In the simulated sunlight irradiation test, since the light having a wavelength lower than 420 nm was filtered, the catalytic efficiency of all the samples was reduced to different degrees (Fig. 6.17b), and the results are summarized in Table 6.1. When the CSs doping amount is less than 1% or is flat, a small amount of electrons can be

Table 6.1 Summary of the physical and photocatalytic properties of the materials

Sample	BET surface area ($m^2 g^{-1}$)	Pore volume ($cm^3 g^{-1}$)	Pore diameter (nm)	Absorbing boundary (nm)	Bandgap energy (eV)	Adsorption efficiency (%)	g_1 (%)	g_2 (%)
P-BOI	20.5	0.120	26.4	410	3.02	7.6	73.7	59.2
CSs-BOI-0.5	28.2	0.121	18.9	420	2.95	6.4	82.2	61.6
CSs-BOI-1	33.0	0.150	19.3	426	2.91	7.8	92.9	67.2
CSs-BOI-3	39.8	0.177	18.6	456	2.72	8.7	78.8	73.5
CSs-BOI-5	51.8	0.196	15.5	585	2.12	10.8	83.3	83.0
CSs-BOI-10	71.8	0.328	20.7	898	1.38	20.5	82.0	72.6
CSs-BOI-15	57.6	0.378	29.0	/	/	19.4	79.6	70.6
CSs-BOI-25	37.6	0.129	15.8	/	/	16.4	48.9	44.9
CSs	6.2	0.016	20.8	/	/	/	/	/

The Efficiency of gaseous Hg^0 removal under LED light irradiation (g_1), the Efficiency of gaseous Hg^0 removal under simulated sunlight irradiation (g_2)
Reprinted from Ref. [40], Copyright 2017, with permission from Elsevier

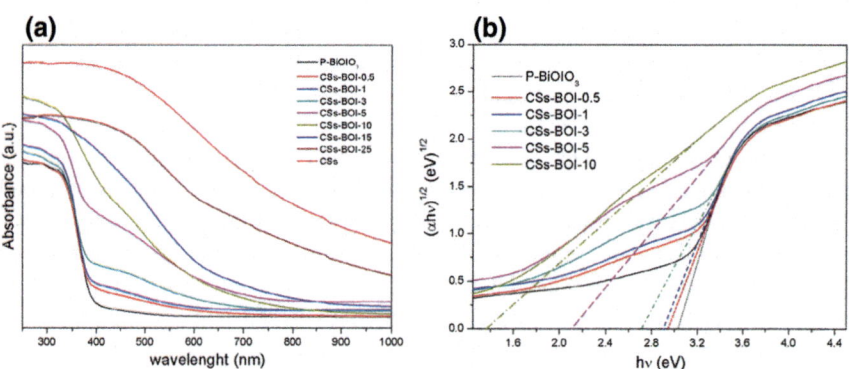

Fig. 6.16 UV–vis diffuse reflectance spectrum (**a**) and the associated (ahm) 1/2 versus (hm) plot (**b**) of the CSs, BiOIO₃ and CSs-BiOI/BiOIO₃. Reprinted from Ref. [40], Copyright 2017, with permission from Elsevier

excited because the bandgap is still too wide (ca. 3.03, 2.95, and 2.91 eV respectively) and the efficiency is reduced. While the CSs doping amount is more than 3%, the bandgap (about 2.72 eV) of the photocatalyst narrows, which is in agreement with the simulated sunlight wavelength. However, oxidation capacity decreased because of the excessive BiOI and CSs content as the upward shift of the valence band. In

6.3 The Photocatalytic Removal of Mercury by Other Photocatalysts 131

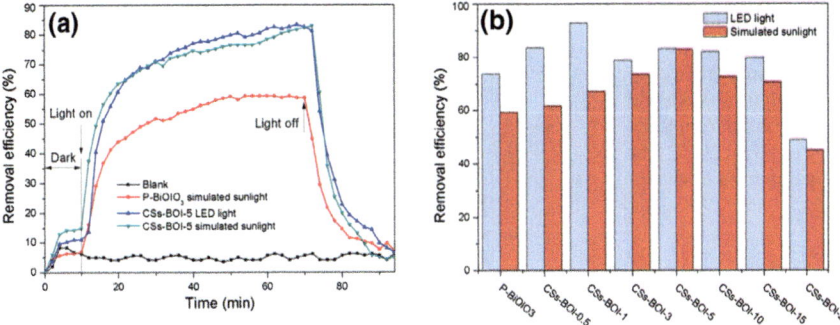

Fig. 6.17 **a** The mercury removal without photocatalysts, over $BiOIO_3$ and CSs-BOI-5, **b** the mercury removal efficiency of $BiOIO_3$, CSs-BiOI/$BiOIO_3$ under different light irradiation. Reprinted from Ref. [40], Copyright 2017, with permission from Elsevier

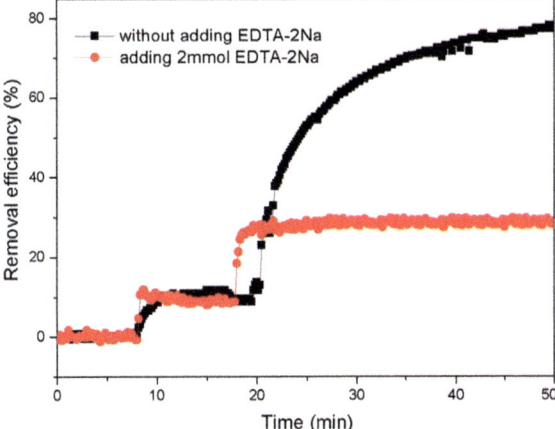

Fig. 6.18 The mercury removal over CSs-BOI-5 in trapping experiment under simulated sunlight irradiation. Reprinted from Ref. [40], Copyright 2017, with permission from Elsevier

addition, the decreasing of BET surface area made photocatalytic activity decrease. In summary, the reduction of the photocatalytic activity resulted in the decrease of the efficiency.

As described previously, Hg^0 cannot be oxidized to Hg^{2+} only with light but no photocatalysts. The trapping experiment was implemented to investigate the major active species generated in the Hg removal process. The scavengers ethylene diaminetetraacetic acid disodium salt (EDTA-Na_2) was added in the photocatalytic system of Hg removal as quenchers of holes (h^+), and the results are shown in Fig. 6.18. Under the same experimental conditions, the removal efficiency of Hg^0 was significantly reduced because of the addition of EDTA-Na2. It proved that photogenerated holes were the main energetic species responsible for the Hg removal. In the presence of photocatalysts, the Hg^0 reacts with the energetic species h^+ with great oxidation ability and turns into Hg^{2+}, involving seven reactions, which can be described as Eqs. (6.58)–(6.65) [43, 44].

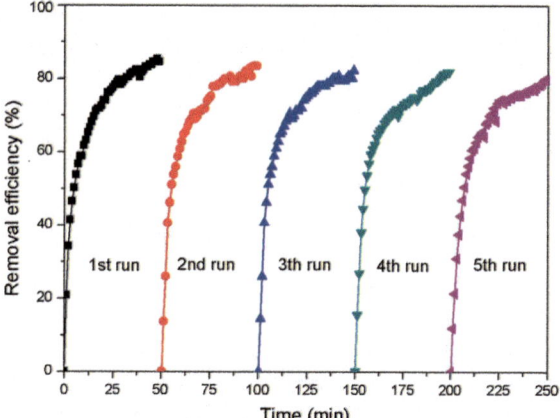

Fig. 6.19 Stability evaluation for the CSs-BOI-5 after five consecutive cycling photocatalytic oxidation of Hg^0. Reprinted from Ref. [40], Copyright 2017, with permission from Elsevier

$$\frac{BiOIO_3}{BiOI}/CSs + h\nu \rightarrow \frac{BiOIO_3}{BiOI}/CSs^*(h^+ + e^-) \quad (6.58)$$

$$H_2O \leftrightarrow H^+ + OH^- \quad (6.59)$$

$$O_2 + 4H^+ + 4e^- \rightarrow 2H_2O \quad (6.60)$$

$$OH_{ad}^- + h^+ \rightarrow \cdot OH_{ad} \quad (6.61)$$

$$H_2O_{ad} + h^+ \rightarrow \cdot OH_{ad} + H^+ \quad (6.62)$$

$$Hg^0(ad) + 2h^+ \rightarrow Hg^{2+} \quad (6.63)$$

$$Hg_{ad}^0 + \cdot OH_{ad} + H^+ \rightarrow Hg^+ + H_2O \quad (6.64)$$

$$Hg_{ad}^+ + \cdot OH_{ad} + H^+ \rightarrow Hg^{2+} + H_2O \quad (6.65)$$

6.3.2.2 Stability of the Photocatalyst

The mercury removal performance of CSs-BiOI/BiOIO$_3$ is excellent, so it is very necessary to study their stability. As shown in Fig. 6.19, the CSs-BOI-5 mounted on a square glass plate was first exposed directly to the simulated sunlight under simulated sunlight conditions, and then the LED was turned off when the concentration of Hg^0 tended to be stable. Then the next cycle started, and it was with a total of 5 cycles. After 5 cycles, the removal rate of Hg^0 by CSs-BOI-5 was about 80% under simulated sunlight irradiation. This shows that CSs-BOI-5 has almost no change during the mercury removal experiment and has good stability. It is a significant and stable photocatalyst for mercury removal from flue gas.

6.3 The Photocatalytic Removal of Mercury by Other Photocatalysts

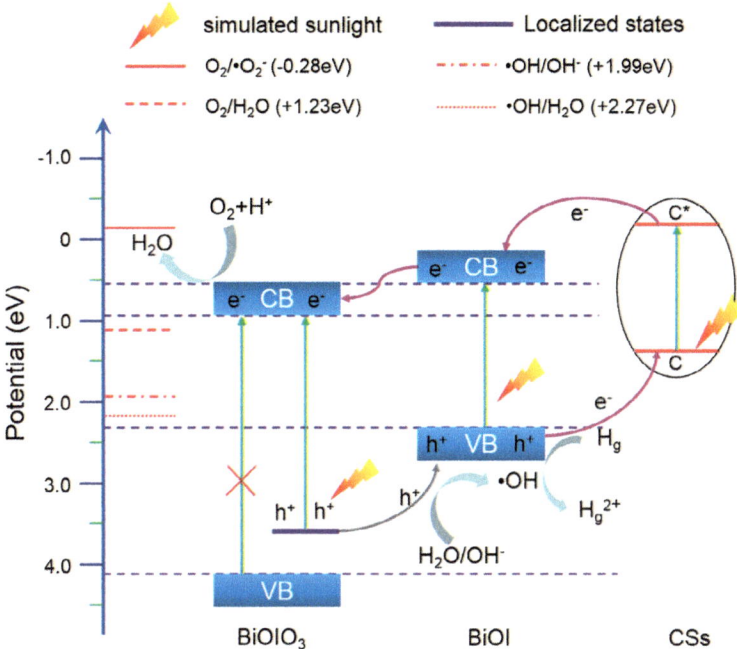

Fig. 6.20 Schematic diagram of photoexcited electron-hole separation process. Reprinted from Ref. [40], Copyright 2017, with permission from Elsevier

6.3.2.3 Photocatalytic Reaction Mechanisms

It is known that, when coupling $BiOIO_3$, BiOI and CSs, a heterojunction photocatalyst between $BiOIO_3$, BiOI and CSs will be formed. Figure 6.20 shows the separation of photoexcited electrons and holes, which is based on the band gap structures of $BiOIO_3$ and BiOI.

After carbon is doped, a localized state can appear on the valence band and the band gap of $BiOIO_3$ is reduced, which causes the response of the light to extend into the visible region. On the other hand, CSs act as photosensitizers, which can absorb radiation and can also be excited to produce electrons and inject electrons into the conduction band of $BiOIO_3$ and BiOI [45]. At the same time, because CS has a positive charge, it may carry away one electron from the valence band, leaving a hole behind. In addition, CSs doping also changed the surface properties of the previous $BiOIO_3$, increased the specific surface area, and in turn increased the Hg^0 adsorption. In summary, the doping of CSs can effectively enhance the photocatalytic reaction. Under light irradiation, because the conduction band position of $BiOIO_3$ is lower than that of BiOI, the excited electrons on the BiOI conduction band can be easily transferred to the conduction band of $BiOIO_3$. However, we find that these addictive electrons in the CB of $BiOIO_3$ cannot reduce $\cdot O_2^-$ to yield O_2, being mainly due to the fact that the latent of the electrons are not negative with respect to $O_2/\cdot O_2^-$

(−0.28 eV) couple. Alternatively, electrons can be consumed by multi-electrons reaction to produce H_2O and O_2/H_2O couple is 1.23 eV [Eq. (6.60)] [46]. At the same time, the active holes on the valence band of $BiOIO_3$ will move to the valence band of BiOI, and the valence band level of BiOI is positive with respect to ·OH/OH⁻ (+1.99 eV) or ·OH/H_2O (+2.27 eV) couple, so the photogenerated holes can change OH⁻ or H_2O into ·OH [Eqs. (6.61) and (6.62)]. Therefore, all factors effectively enhance the separation extent and inhibit the recombination of the photoinduced electron and hole carriers on the CSs-BiOI/$BiOIO_3$. Finally, the hydroxyl radicals are extremely oxidizing, being capable to effectively oxidize Hg^0 into Hg^{2+}. Thus, the CSs-BiOI/$BiOIO_3$ indicates an efficient visible light photocatalytic activity in Hg^0 removal performances.

6.3.3 The Photocatalytic Removal of Mercury by ZnO

In this section, CuO / ZnO metal oxide was prepared by homogeneous coprecipitation using zinc acetate and copper acetate as original materials. The photocatalytic performance of CuO / ZnO metal oxides in the removal of Hg^0 by photocatalytic oxidation was studied in coal-fired flue gas.

6.3.3.1 XRD Analysis

Figure 6.21 is an X-ray diffraction spectrum of CuO, ZnO, 0.01-CuO / ZnO, 0.02-CuO / ZnO, 0.1-CuO / ZnO, 0.2-CuO / ZnO, 1-CuO / ZnO, and 2-CuO / ZnO photocatalyst. The spectrum shows that the finally prepared sample is a CuO / ZnO mixed metal oxide. Compared with the standard card, CuO is monoclinic (JCPDS file no. 89-5895) and the diffraction peak of ZnO fits nicely with the standard peak of hexagonal ZnO (JCPDS file no. 36-1451). The (110) diffraction peak ($2\theta \approx 32.73$) of CuO shifts toward a small angle in CuO/ZnO. As the molar ratio increases, the shift angle first enhances and then decreases, and when the molar ratio is 0.2, it shows the maximum offset angle.

6.3.3.2 UV-vis DRS

Figure 6.22 is UV-vis absorption spectrum of pure ZnO, CuO, CuO/ZnO metal oxide. It can be observed that the characteristic assimilation boundary of pure ZnO is about 400 nm, and the characteristic assimilation boundary of CuO is broader. The photoabsorption intensity of CuO-doped ZnO increased with the increase of doping molar ratio, and its photosensitivity increased in the visible region, indicating red shift of ZnO doped with CuO in response to the visible region [47]. This may be due to the formation of CuO / ZnO heterostructures, which reduces the Fermi level of the metal

6.3 The Photocatalytic Removal of Mercury by Other Photocatalysts

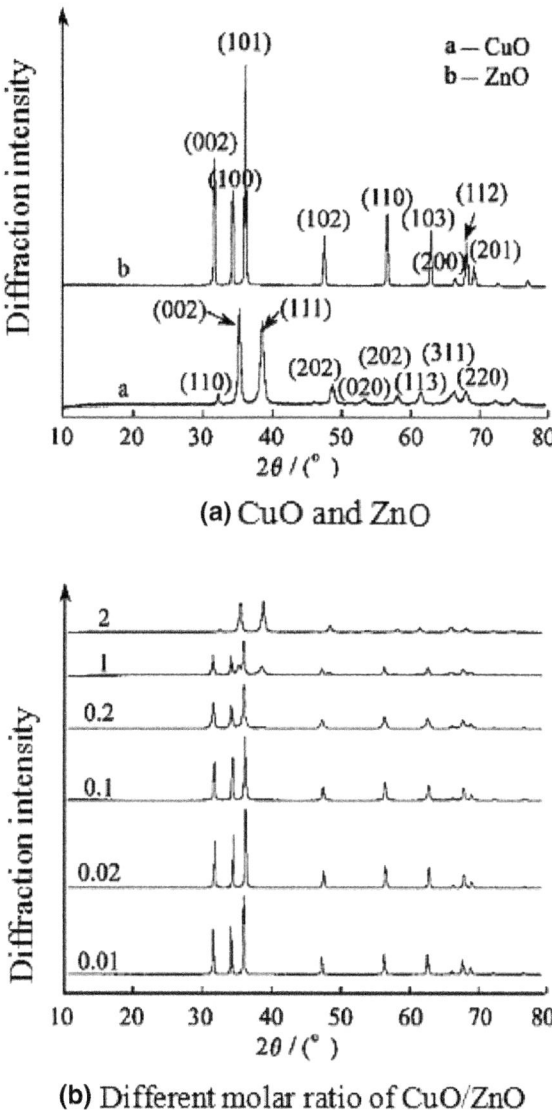

Fig. 6.21 X-ray diffraction patterns of different photocatalysts

oxides, inhibits the recombination of the optical carriers and in turn enhances the photocatalytic performance.

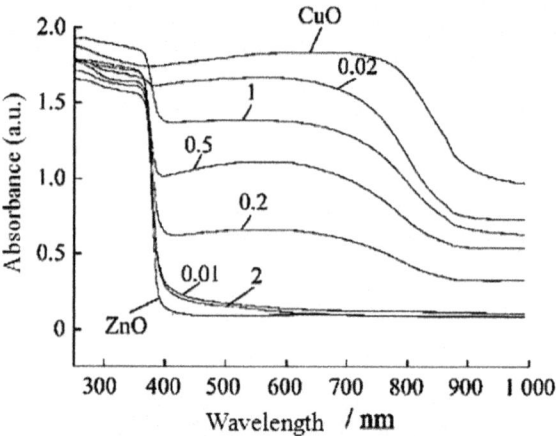

Fig. 6.22 UV-Vis absorption spectra of different samples

6.3.3.3 BET and N_2 Adsorption-Desorption Isotherm

The concrete surface area of different samples was surveyed by Micrograph Ultra High Performance Automatic Gas Adsorption Analyzer ASAP 2020 HD88. The specific surface area relatively increased when pure ZnO was doped CuO and at a molar ratio of 0.2, the surface area is the largest. This difference may be resulted by the structure of CuO/ZnO metal oxide microporous structure. The specific surface area increases, the related area between the flue gas and the photocatalyst increases. The larger specific surface area improves photocatalytic reaction efficiency of catalyst [48], which is consistent with the final experimental results.

6.3.3.4 Photocatalytic Properties

Pure CuO and pure ZnO have mercury removal effect, which is mainly because of its excitation of electrons and holes in light conditions [49]. Water vapor in the air is decomposed into H^+ and OH^-. Under the action of photo-generated holes, OH^- formed ·OH with strong oxidizing ability. Under the action of photogenic electrons, O_2 formed $·O_2^-$ with strong oxidizing ability. Both ·OH and $·O_2^-$ can oxidize elemental mercury to divalent mercury [50]. ZnO is a wide-band-gap semiconductors (3.4 eV), which is low quantum yield and low photocatalytic activity. The forbidden band width of CuO is narrow (1.2 eV), which is strong photosensitivity and easily excited electrons and holes under light. However, a narrow band gap makes the excited electrons and holes recombine quickly, which greatly reduces the photocatalytic effect. Therefore, we examined the mercury removal performance of the modified ZnO under UV light conditions, as shown in Fig. 6.23.

As can be seen from Fig. 6.23, with the conditions of UV light, the photocatalytic removal efficiency of mercury is 15.47, 12.98, 14.48, 34.19, 14.36, 57.4, 18.53, 12.45

6.3 The Photocatalytic Removal of Mercury by Other Photocatalysts

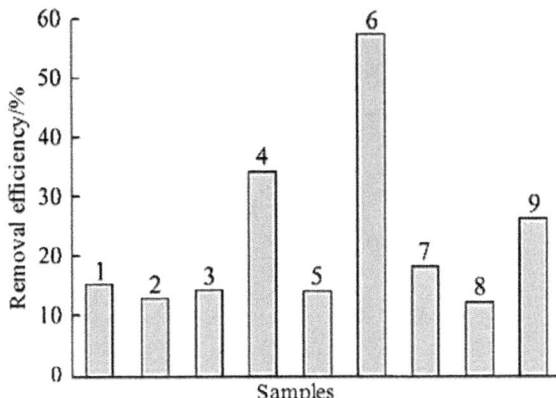

Fig. 6.23 Mercury removal efficiency of different molar ratio samples: 1-ZnO; 2-CuO; 3-0.01-ZnO /CuO; 4-0.02-ZnO/CuO; 5-0.1-ZnO/CuO; 6-0.2-ZnO/CuO; 7-0.5-ZnO/CuO; 8-1-ZnO/CuO; 9-2-ZnO/CuO

and 26.54% respectively. When the molar ratio of CuO/ZnO is 0.2, the mercury removal efficiency is best.

The modified ZnO greatly enhances the photocatalytic activity, mainly because CuO/ZnO forms a heterojunction structure on the surface of the nanocomposite, which impedes the recombination of photogenerated electrons and holes and enhances the use of the optical carriers [51, 52]. Photogenerated electrons migrate from conduction band of high-potential ZnO to conduction band of low-potential CuO, while photogenerated holes migrate from VB of CuO to VB of ZnO, thus it largely inhibits the recombination of the optical carriers and increases the photocatalytic efficiency.

The photocatalytic reaction equation is as follows:

$$CuO/ZnO + h\upsilon \rightarrow CuO/ZnO + h^+ + e^- \quad (6.66)$$

$$H_2O \leftrightarrow H^+ + OH^- \quad (6.67)$$

$$OH^-_{ad} + h^+ \rightarrow \cdot OH_{ad} \quad (6.68)$$

$$H_2O_{ad} + h^+ \rightarrow \cdot OH_{ad} + H^+ \quad (6.69)$$

$$O_{2ad} + e^- \rightarrow O^-_{2ad} \quad (6.70)$$

$$O_{2ad} + H^+ \rightarrow \cdot HO_{2ad} \quad (6.71)$$

$$\cdot HO_{2ad} + e^- + H^+ \rightarrow H_2O_{2ad} \quad (6.72)$$

$$H_2O_{2ad} + h_\upsilon \rightarrow 2 \cdot OH_{ad} \quad (6.73)$$

$$2Hg^0_{ad} + O^-_{2ad} \rightarrow HgO_{ad} \quad (6.74)$$

$$Hg^0_{ad} + 2 \cdot OH_{ad} \rightarrow HgO_{ad} + H_2O \quad (6.75)$$

References

1. J. Wu, C. Li, X. Zhao, Q. Wu, X. Qi, X. Chen, T. Hu, Y. Cao, Photocatalytic oxidation of gas-phase Hg^0 by CuO/titanium dioxide. Appl. Catal. B: Environ. **176–177**, 559–569 (2015)
2. W.K. Wang, J.J. Chen, M. Gao, Y.X. Huang, X. Zhang, H.Q. Yu, Photocatalytic degradation of Atrazine by boron-doped titanium dioxide with a tunable rutile/anatase ratio. Appl. Catal. B Environ. **195**, 69–76 (2016)
3. X. Zhou, J. Wu, Q. Li, Y. Qi, Z. Ji, P. He, X. Qi, P. Sheng, Q. Li, J. Ren, Improved electron-hole separation and migration in V_2O_5/rutile-anatase photocatalyst system with homo-hetero junctions and its enhanced photocatalytic performance. Chem. Eng. J. **330**, 294–308 (2017)
4. C. He, B. Shen, J. Chen, J. Cai, Adsorption and oxidation of elemental mercury over Ce-MnO_x/Ti-PILCs. Environ. Sci. Technol. **48**(14), 7891–7898 (2014)
5. S. Yang, Y. Guo, N. Yan, D. Wu, H. He, J. Xie, Z. Qu, J. Jia, Remarkable effect of the incorporation of titanium on the catalytic activity and SO_2 poisoning resistance of magnetic Mn-Fe spinel for elemental mercury capture. Appl. Catal. B Environ. **101**, 698–708 (2011)
6. H. Huang, Y. He, X. Li, M. Li, C. Zeng, F. Dong, X. Du, T. Zhang, Y. Zhang, BiO(OH)(NO) as a desirable [Bi O] layered photocatalyst: strong intrinsic polarity, rational band structure and 001 active facets co-beneficial for robust photooxidation capability. J. Mater. Chem. A **3**, 24547–24556 (2015)
7. H. Huang, X. Li, J. Wang, F. Dong, P.K. Chu, T. Zhang, Y. Zhang, Anionic group self-doping as a promising strategy: band-gap engineering and multi-functional applications of high-performance CO32-doped Bi2O2CO3. ACS Catal. **5**, 4094–4103 (2015)
8. N. Fernández, M.A. Lopezanton, M. Díaz-Somoano, M.R. Rosa, Effect of oxy-combustion flue gas on mercury oxidation. Environ. Sci. Technol. **48**, 7164–7170 (2014)
9. A. Zhang, L. Zhang, H. Lu, G. Chen, Z. Liu, J. Xiang, L. Sun, Facile synthesis of ternary Ag/AgBr-Ag_2CO_3 hybrids with enhanced photocatalytic removal of elemental mercury driven by visible light. J. Hazardous Mater. **314**, 78–87 (2016)
10. K. Vijayarangamuthu, E. Han, K.J. Jeon, Low frequency ultrasonication of Degussa P25 titanium dioxide and its superior photocatalytic properties. J. Nanosci. Nanotechnol. **16**, 4399–4404 (2016)
11. Y. Hong, Y. Jiang, C. Li, W. Fan, Y. Xu, M. Yan, W. Shi, In-situ synthesis of direct solid-state Z-scheme V_2O_5/g-C_3N_4 heterojunctions with enhanced visible light efficiency in photocatalytic degradation of pollutants. Appl. Catal. B Environ. **180**, 663–673 (2016)
12. Y. Wang, Y.R. Su, L. Qiao, L.X. Liu, Q. Su, C.Q. Zhu, X.Q. Liu, Synthesis of one-dimensional titanium dioxide/V_2O_5 branched heterostructures and their visible light photocatalytic activity towards Rhodamine B. Nanotechnology **22**, 225702–225708 (2011)
13. G. Xiong, R. Shao, T.C. Droubay, A.G. Joly, K.M. Beck, S.A. Chambers, W.P. Hess, Photoemission electron microscopy of titanium dioxide Anatase films embedded with rutile nanocrystals. Adv. Funct. Mater. **17**, 2133–2138 (2007)
14. L. Zhao, C. Li, Y. Wang, H. Wu, L. Gao, J. Zhang, G. Zeng, Simultaneous removal of elemental mercury and NO from simulated flue gas using a CeO_2 modified V_2O_5-WO_3/titanium dioxide catalyst. Catal. Sci. Techn. **6**, 420–430 (2016)
15. W.-T. Chen, V. Jovic, D. Sun-Waterhouse, H. Idriss, G.I.N. Waterhouse, The role of CuO in promoting photocatalytic hydrogen production over titanium dioxide. Int. J. Hydrogen Energy **38**, 15036–15048 (2013)
16. Y. Yuan, J. Zhang, H. Li, Y. Li, Y. Zhao, C. Zheng, Simultaneous removal of SO_2, NO and mercury using titanium dioxide-aluminum silicate fiber by photocatalysis. Chem. Eng. J. **192**, 21–28 (2012)
17. Masato Takeuchi, Satoru Dohshi, Takashi Eura, Masakazu Anpo, Preparation of titanium—silicon binary oxide thin film photocatalysts by an ionized cluster beam deposition method. Their photocatalytic activity and photoinduced super-hydrophilicity. ChemInform **35**(10), 14278–1428 (2004)

18. Y. Jiang, P. Zhang, Z. Liu, F. Xu, The preparation of porous nano-titanium dioxide with high activity and the discussion of the cooperation photocatalysis mechanism. Mater. Chem. Phys. **99**, 498–504 (2006)
19. Z.H. Wang, S.D. Jiang, Y.Q. Zhu, J.S. Zhou, J.H. Zhou, Z.S. Li, K.F. Cen, Investigation on elemental mercury oxidation mechanism by non-thermal plasma treatment. Fuel Process. Technol. **91**, 1395–1400 (2010)
20. J. Wu, C. Li, X. Chen, J. Zhang, L. Zhao, T. Huang, T. Hu, C. Zhang, B. Ni, X. Zhou, P. Liang, W. Zhang, Photocatalytic oxidation of gas-phase Hg^0 by carbon spheres supported visible-light-driven CuO-titanium dioxide. J. Ind. Eng. Chem. **46**, 416–425 (2017)
21. M.A. Mohamed, N.W.S. Wan, J. Jaafar, M.S. Rosmi, Z.A.M. Hir, M.A. Mutalib, A.F. Ismail, M. Tanemura, Carbon as amorphous shell and interstitial dopant in mesoporous rutile titanium dioxide: bio-template assisted sol-gel synthesis and photocatalytic activity. Appl. Surf. Sci. **393**, 46–59 (2017)
22. P. Zhang, C. Shao, Z. Zhang, M. Zhang, J. Mu, Z. Guo, Y. Liu, $TiO_{(2)}$@carbon core/shell nanofibers: controllable preparation and enhanced visible photocatalytic properties. Nanoscale **3**, 2943–2949 (2011)
23. X. Wu, S. Yin, Q. Dong, C. Guo, H. Li, T. Kimura, T. Sato, Synthesis of high visible light active carbon doped titanium dioxide photocatalyst by a facile calcination assisted solvothermal method. Appl. Catal. B **142–143**, 450–457 (2013)
24. C.C. Wang, J.R. Li, X.L. Lv, Y.Q. Zhang, G. Guo, Photocatalytic organic pollutants degradation in metal-organic frameworks. Energy Environ. Sci. **7**, 2831–2867 (2014)
25. X. Zhou, J. Wu, Q. Li, T. Zeng, Z. Ji, P. He, W. Pan, X. Qi, C. Wang, Pankun Liang, Carbon decorated In_2O_3 /titanium dioxide heterostructures with enhanced visible-light-driven photocatalytic activity. J. Catal. **355**, 26–39 (2017)
26. J. Jia, D. Li, J. Wan, X. Yu, Characterization and mechanism analysis of graphite/C-doped titanium dioxide composite for enhanced photocatalytic performance. J. Indust. Engineer. Chem. **33**, 162–169 (2016)
27. G. Liu, H.G. Yang, J. Pan, Y.Q. Yang, G.Q. Lu, H.M. Cheng, Titanium dioxide crystals with tailored facets. Chemistry Rev. **114**, 9559–9612 (2014)
28. J.B. Mu, B. Chen, M.Y. Zhang, Z.C. Guo, P. Zhang, Z.Y. Zhang, Y.Y. Sun, C.L. Shao, Y.C. Liu, Enhancement of the visible-light photocatalytic activity of In_2O_3-titanium dioxide nanofiber heteroarchitectures. ACS Appl. Mater. Interfaces. **4**, 424–430 (2012)
29. R. Marschall, Photocatalysis: Semiconductor composites: Strategies for enhancing charge carrier separation to improve photocatalytic activity. Adv. Funct. Mater. **24**, 2421–2440 (2014)
30. L. Zhao, C. Li, Y. Wang, H. Wu, L. Gao, J. Zhang, G. Zeng, Simultaneous removal of elemental mercury and NO in simulated flue gas over V_2O_5 /ZrO_2-CeO_2 catalyst. Catal. Sci. Technol. **6**, 420–430 (2016)
31. Y. Cao, Q. Li, C. Li, J. Li, J. Yang, Surface heterojunction between (001) and (101) facets of ultrafine anatase titanium dioxide nanocrystals for highly efficient photoreduction CO_2 to CH_4. Appl. Catal. B **198**, 378–388 (2016)
32. R. Zhou, J. Wu, J. Zhang, H. Tian, P. Liang, T. Zeng, P. Lu, J. Ren, T. Huang, X. Zhou, P. Sheng, Photocatalytic oxidation of gas-phase Hg^0 on the exposed reactive facets of $BiOI/BiOIO_3$ heterostructures. Appl. Catal. B **204**, 465–474 (2017)
33. Y. Su, L. Zhang, W.Z. Wang, Internal polar field enhanced H_2 evolution of $BiOIO_3$ nanoplates. Int. J. Hydrogen Energy **41**, 10170–10177 (2016)
34. L.B. Hou, S. Li, Y.H. Lin, D.J. Wang, T.F. Xie, Photogenerated charges transfer across the interface between NiO and titanium dioxide nanotube arrays for photocatalytic degradation: a surface photovoltage study. J. Colloid Interface Sci. **465**, 96–102 (2016)
35. L.P. Jiang, H.L. Yu, L.Y. Shi et al., Optical band structure and photogenerated carriers transfer dynamics in FTO/titanium dioxide heterojunction photocatalysts. Appl. Catal. B **199**, 224–229 (2016)
36. M.Y. Guo, A.M.C. Ng, F.Z. Liu et al., Photocatalytic activity of metal oxides—the role of holes and $^{\cdot}$OH radicals. Appl. Catal. B **107**, 150–157 (2011)

37. Y. Yuan, Y. Zhao, H. Li, Y. Li, Electrospun metal oxide-titanium dioxide nanofibers for elemental mercury removal from flue gas. J. Hazard. Mater. **227–228**, 427–435 (2012)
38. K.J. Lee, C.W. Choi, W. Platinized, WO_3 as an environmental photocatalyst that generates ·OH radicals under visible light. Environ. Sci. Technol. **44**, 6849–6854 (2010)
39. Y. Yuan, J. Zhang, H. Li, Y. Li, Y. Zhao, C. Zheng, Simultaneous removal of SO_2, NO and mercury using titanium dioxide -aluminum silicate fiber by photocatalysis. Chem. Eng. J. **192**, 21–28 (2012)
40. J. Wu, X. Chen, C. Li, Y. Qi, X. Qi, J. Ren, B. Yuan, B. Ni, R. Zhou, J. Zhang, T. Huang, Hydrothermal synthesis of carbon spheres–BiOI/$BiOIO_3$ heterojunctions for photocatalytic removal of gaseous Hg^0 under visible light. Chem. Eng. J. **304**, 533–543 (2016)
41. Y. Huang, W. Ho, S. Lee, L. Zhang, G. Li, J.C. Yu, Effect of carbon doping on the mesoporous structure of nanocrystalline titanium dioxide and its solar-light-driven photocatalytic degradation of NO_x. Langmuir **24**, 3510–3516 (2008)
42. L. Jiang, Y. Huang, T. Liu, Enhanced visible-light photocatalytic performance of electrospun carbon-doped titanium dioxide/halloysite nanotube hybrid nanofibers. J. Colloid Interface Sci. **439**, 62–68 (2015)
43. J. Kim, C.W. Lee, W. Choi, Platinized WO_3 as an environmental photocatalyst that generates OH radicals under visible light. Environ. Sci. Technol. **44**, 6849–6854 (2010)
44. Y. Yuan, J. Zhang, H. Li, Y. Li, Y. Zhao, C. Zheng, Simultaneous removal of SO_2, NO and mercury using titanium dioxide-aluminum silicate fiber by photocatalysis. Chem. Eng. J. **192**, 21–28 (2012)
45. Q. Chen, Q. Wu, Fabrication of carbon microspheres@ $PbMoO_4$ core-shell hybrid structures and its visible light-induced photocatalytic activity. Catal. Commun. **24**, 85–89 (2012)
46. A. Kudo, Y. Miseki, Heterogeneous photocatalyst materials for water splitting. Chem. Soc. Rev. **38**, 253–278 (2009)
47. Bu Ni, Li Chaoen, Zhao Lili, Zhang Jing, Zhao Zhen, YVO_4 nano photocatalyst preparation and photocatalytic performance evaluation of mercury removal. J. Shanghai Univ. Electr. Power **5**, 434–438 (2015)
48. Z. Wen, Z. Hua Ji, P. Yi Ru, Modification technology of titanium dioxide semiconductor photocatalyst. J. Fujian Normal Univ. Nat. Sci. Ed. **4**, 113–123 (2005)
49. Y. Yuan, J. Zhang, H. Li, Y. Li, Y. Zhao, Simultaneous removal of SO_2, NO and mercury using titanium dioxide-aluminum silicate fiber by photocatalysis. Chem. Eng. J. **92**, 21–28 (2012)
50. S. Sjostrom, M. Durham, C.J. Bustard, C. Martin, Activated carbon injection for mercury control: Overview. Fuel **89**(6), 1320–1322 (2010)
51. Y. Jiang, P. Zhang, Z. Liu, F. Xu, The preparation of porous nano-titanium dioxide with high activity and the discussion of the cooperation photocatalysis mechanism. Mater. Chem. Phys. **99**(2–3), 498–504 (2004)
52. Z.H. Wang, S.D. Jiang, Y.Q. Zhu, J.S. Zhou, J.H. Zhou, Z.S. Li, K.F. Cen, Investigation on elemental mercury oxidation mechanism by non-thermal plasma treatment. Fuel Process. Technol. **91**(11), 1395–1400 (2010)

Chapter 7
The Photocatalytic Technology for Wastewater Treatment

Abstract In order to maintain the balance of desulfurization slurry material circulation system, prevent the soluble part of the chlorine concentration in flue gas to exceed a specified value, and ensure the quality of gypsum, discharge from the system is necessary, including a certain amount of waste water, which is mainly from gypsum dehydration and cleaning system. The impurities in waste water mainly include suspended, supersaturated sulfites, sulfates, and heavy metals, many of which are the primary pollutants required to be strictly controlled by national environmental standards. The photocatalysis showed great superiority on wastewater treatment. Based on the photocatalytic mechanism and the kinetics, the photocatalytic process includes primary reaction process and secondary reaction process, and the wastewater degraded with photocatalysts is studied in this chapter. Through the analysis of traditional method on degrading wastewater, it is a universal view that photocatalysis is a promising method with industrialized value. At last, several future developing views of the application of photocatalysts on wastewater treatment were put forward, which included efficiency priority, combining immobilization, mechanism in microcosmic, and application in macroscopic.

Keywords Photocatalytic · Wastewater · Water treatment · Purification

7.1 The Principle of Photocatalytic Wastewater Treatment

In the past few decades, advanced oxidation processes (AOPs) have been applied to remove persistent organic contaminants in water [1–3]. There are some disadvantages in simplex ozonation or photocatalysis related to AOPs which restrain their efficiency in the elimination of organic pollutants. Because of low oxidation rate, photocatalysis is a sluggish process [4], while ozonation alone often results in incomplete mineralization of organic compounds [5]. These shortcomings make the practical application of these individual technologies to economically dispose industrial wastewater undesirable. To solve these problems, the presentation of catalysts and/or light irradiation to the wastewater during ozonation is urgent to promote the elimination efficiency.

The association of photocatalysis and ozonation has been reported as a likely and eco-friendly treatment approach for organic pollutants [6, 7], because this coupled technique is able to achieve high removal efficiency and does not reproduce any secondary pollution. Ozone, as a formidable scavenger, can quickly catch the photogenerated electrons when photocatalysis and ozonation processes are conducted at the same time. This will promote the transport of photoexcited charge carriers and the utilization of ozone, therefore, producing more strong species such as h^+, $\cdot OH$ and $\cdot O^{2-}$ to speed up the decomposition and mineralization of organic pollutants [8–12]. Some semiconductor materials have been reported to spark a synergistic effect between photocatalysis and ozonation to greatly enhance the oxidation efficiency of organic contaminants [13, 14]. The $\cdot OH$ has a reaction energy of 402 MJ/mol and can destroy C–C, C–H, C–O, C–N, N–H bonds in organic matter, oxidizing refractory organisms to inorganic substances such as CO_2 and H_2O.

The titanium dioxide has unique properties. It is easy to prepare, inexpensive, non-toxic, and has good stability and photoreactivity. It has been widely used in life as a catalyst for photocatalytic ozonation of organic substances [15–18]. However, the use of pure titanium dioxide as a photocatalytic ozonation catalyst also has some drawbacks, such as the rapid reorganization of light-induced electron holes and the incomplete mineralization of organic matter, which will to some extent make it unable to be used in real life to achieve the treatment of wastewater [12, 19, 20]. Therefore, many effective methods have been taken to improve the catalytic activity of titanium dioxide.

The coupling titanium dioxide with transition metal oxide to construct composites such as ZnO/titanium dioxide [21], Cu_2O/titanium dioxide [22], WO_3/titanium dioxide [23, 24], ZrO/titanium dioxide [25], and MoO_3/titanium dioxide [26, 27], significantly enhance the photocatalytic activity as a result of the accelerated transfer of photoinduced charge carriers and more surface hydroxyl groups. It is generally regarded that the abundance of surface hydroxyl groups of the catalyst could also enhance degradation rate and mineralization efficiency of organics in ozonation process [28].

7.2 Titanium-Based Photocatalysts

7.2.1 Pure Titanium Dioxide

In 1972, Japanese scholar Fujishima and Honda reported the photocatalytic degradation of H_2 and O_2 on n-type semiconductor titanium dioxide single crystal electrode. This report makes the semiconductor photocatalytic redox technology in sewage treatment, anti-bacterial and anti-virus and other potential applications be received widespread attention, which has been rapidly evolving.

Dye wastewater contains benzene ring, amine, azo groups and other carcinogens, while the degradation efficiency of conventional methods to deal with water-soluble dyes is usually low. The results [29–32] show that the decoloration effect of titanium dioxide on azo dyes, anthraquinone dyes, triarylmethanes and cyanines can reach more than 95%, and the COD removal efficiency is between 80 and 100%. In the treatment of pharmaceutical wastewater, the composition of pharmaceutical wastewater is complex, and the concentration of pollutants is high, which contains difficult-to-degrade substances and antibiotics with antibacterial activity, being toxic and harmful. It belongs to refractory industrial wastewater. The use of photocatalytic oxidation of pharmaceutical wastewater does not generate other toxic substances, no secondary pollution, with incomparable advantages compared to other methods. As for oily wastewater, the degradation rate of oil-containing wastewater by nano-titanium dioxide powder can reach 94.74%, which can effectively treat oils that are insoluble in water and have lower density than water [33].

Because degradation reaction mechanism of organic matter with the photocatalytic oxidation by titanium dioxide is essentially a free radical reaction [34], increasing the photo-generated electron-hole separation efficiency, i.e., suppressing the electron-hole recombination, can increase the photocatalyst intrinsic quantum efficiency and in turn improve the photocatalytic performance of titanium dioxide. Therefore, there are some ways, including the transition metal ions doped titanium dioxide, precious metal surface deposition, semiconductor recombination, the modified photocatalyst for a specific response, to improve the photocatalytic ability of the photocatalyst, which help to improve the intrinsic quantum efficiency of photocatalysts. Then, the catalytic quantum efficiency or visible light utilization could be significantly improved [35].

7.2.2 Tungsten-Doped Titanium Dioxide

The photocatalyst of tungsten (VI)-doped titanium dioxide powder by sol-gel means can be used for the bleaching of crystal violet (CV) under UV radiation in the presence of oxygen. The doping of W(VI) not only changes the surface performances of titanium dioxide, but also alters the overall properties including photocatalytic reactivity. The stability and specific surface area of the anatase phase are enhanced during the heat treatment. The existence of W(VI) on the surface of the material reduces the potential of zero charge (pzc) of the particles, whereas the CV adsorption containment depends mainly on the surface area and the pH, independent of the W(VI) content. The highest bleach rate per gram of photocatalyst is the result of competition between the surface area and the intrinsic bleaching activity at each adsorption site with a W content of up to about 3%. At pH 4.0, doped particles share a collaborative photodegradation mechanism, and N-demethylation has a lower efficiency (in relation to aromatic structure rupture) than that observed over pure titanium dioxide. The reaction mechanism also depends on the pH of the suspension, and N-demethylation is more favorable at higher pH values. The experimental results show that the doping of transition metal cations on titanium dioxide and the change of reaction mechanism by controlling the pH value are of great significance for the purpose of realizing more undesired aromatic structures by specific cleavage.

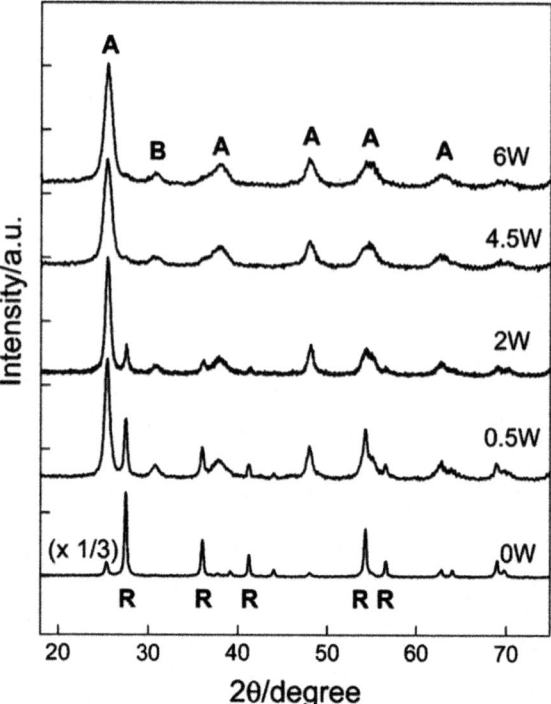

Fig. 7.1 PXRD patterns of titanium dioxide composites with different amount of doped W. Reprinted from Ref. [36], copyright 2018, with permission from ACS

The crystalline structure of titanium dioxide with different W content was characterized by Polycrystalline X-ray diffractometer (PXRD) [36]. Figure 7.1 shows the PXRD pattern of samples loaded with different amounts of tungsten, annealed at 500 °C for 3 h. Sample 0 W displays the coexistence of rutile and anatase. When tungsten is incorporated into titanium dioxide, the above phases co-exist with Brookite, as evidenced by a reflection centered at 31°. The PXRD pattern shows that with the increase of tungsten content, the anatase of the sample becomes richer and the diffraction peak broadens, indicating that the crystal size is smaller.

Figure 7.2 shows that the surface area and isoelectric point of samples are with different amounts of tungsten as determined by BET analysis. Consistent with PXRD results, the surface area increases with increasing tungsten content. The reduction in pzc indicates that a portion of tungsten is doped on the surface. Although both BET surface area and pzc are affected by the percentage of W, the surface area appears to be more affected by the amount of tungsten than pzc, namely the surface area is more sensitive to the amount of tungsten than pzc. The chemical stability of the as-prepared photocatalysts was evaluated by SEM-EDS testing, and the unchanged W/Ti ratio before and after the photocatalytic experiment indicates that these binary oxides are stable to W (VI) lixiviation.

7.2 Titanium-Based Photocatalysts

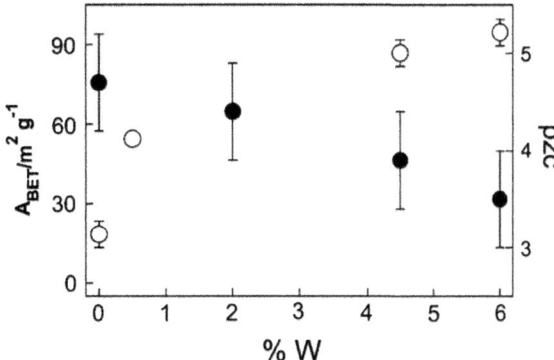

Fig. 7.2 BET surface area (O) and point of zero charge (b) as a function of tungsten content. Reprinted from Ref. [36], copyright 2018, with permission from ACS

7.2.3 Ag⁺-Doped Titanium Dioxide

The two methods, injection and spreading based on a simple solvent-casting method, are used to immobilize floating silver ion-doped titanium dioxide particles in a polystyrene matrix and on a polystyrene matrix. The cost of this photocatalyst doped with floating silver ions is significantly reduced as a result of the elimination of the need of any expensive titanium precursor or high temperature, utilizing inexpensive polymers that are readily available, and mixing the ease of manufacture with high efficiency. The maximum PCAs achieved by immersion and spreading of the doped photocatalyst under UV illumination is about 86% and about 94% respectively. For sunlight, the maximum photo-catalytic activity (PCA) reached 68 and 83% respectively. Due to the doping of titanium dioxide and floating silver ions, this new type of photocatalyst can have higher light utilization and better oxidation. In addition, the high efficiency and stability of such photocatalysts make the bounteously available polystyrene (PS) waste an attractive and suitable substrate to be explored for the fabrication of a low-cost titanium dioxide photocatalyst for environmental remediation.

The XRD patterns of titanium dioxide Degussa P25, Ag⁺-doped titanium dioxide, PSPC (Ag⁺-I-10), and PSPC(Ag⁺-S-10) photocatalysts are shown in Fig. 7.3. The peaks obtained at 2θ of 25.30°, 48.03°, 53.89°, 55.06°, and 62.69° in the case of titanium dioxide Degussa P25 powder, 25.43°, 48.21°, 54.05°, 55.21°, and 62.92° in the case of Ag⁺-doped titanium dioxide P25, 25.17°, 47.99°, 54.05°, 55.03°, and 62.75° for a PSPC(Ag⁺-I-10) sheet, and 25.15°, 47.87°, 53.81°, 54.99°, and 62.57° for a PSPC(Ag⁺-S-10) sheet illustrate that all of the photocatalysts have anatase crystal structures.

Figure 7.4 shows the Rietveld fittings of pure titanium dioxide and titanium dioxide-doped powders. The Ag⁺-titanium dioxide powder and the synthesized doped sheet do not have any diffraction peak characteristics for the doping element Ag and/or its oxide (Ag_2O, AgO). This is probably due to the very low Ag content (1 mol%) in the developed photocatalyst, which is below the X-ray detection limitation. Since Ag⁺ has a higher ion size (1.26 Å) compared to Ti (0.605 Å), Ag⁺ does not incorporate into the lattice of titanium dioxide. There was no major alter in both

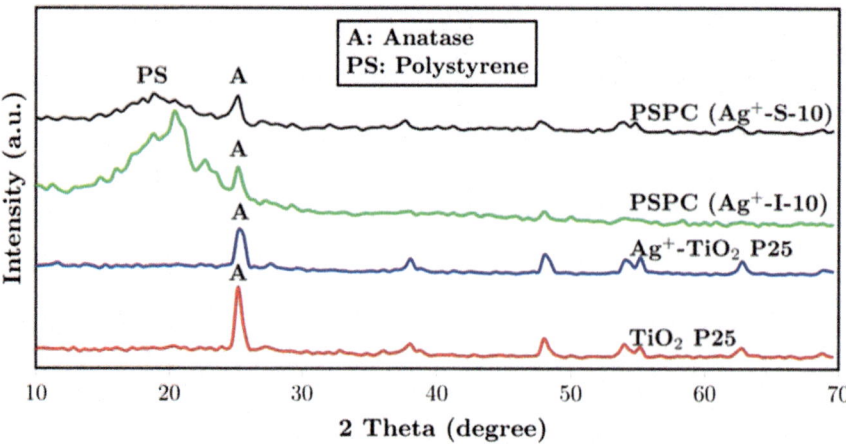

Fig. 7.3 XRD patterns of the titanium dioxide P25, Ag^+-titanium dioxide P25, PSPC(Ag^+-I-10), and PSPC(Ag^+-S-10) photocatalysts. Reprinted from Ref. [37], copyright 2018, with permission from ACS

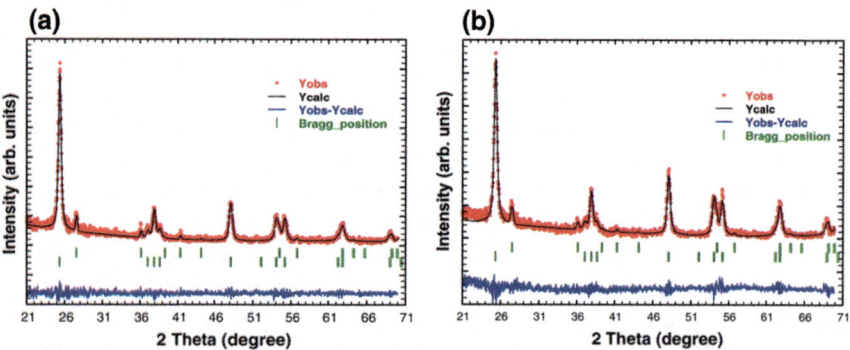

Fig. 7.4 Rietveld refinement of XRD patterns: **a** pure titanium dioxide P25 powder; **b** Ag^+-titanium dioxide P25 powder. Reprinted from Ref. [37], copyright 2018, with permission from ACS

the lattice parameters of undoped and silver-doped titanium dioxide powders and the cell volume, indicating that Ag may not be doped into the titanium dioxide lattice. Therefore, the XRD patterns showed that the crystal structure of titanium dioxide did not change significantly throughout the preparation of the photocatalyst.

7.3 Other Photocatalysts Used in Wastewater Treatment

7.3.1 Zinc-Based Photocatalysts

ZnO has been recently reported as an effective catalyst for ozonation of organic contaminants in aqueous solution [38, 39]. Following this line of thinking, Zinc-based photocatalysts have also been studied.

As potential photocatalytic agents, zinc sulfide (ZnS), silver sulfide (Ag_2S) and bimetallic ZnS-Ag_2S are composed from a single source forerunner and evaluated. The prepared nanoparticles were found to be spherical, crystalline, with a clear atomic plane and a size in the range of 6–12 nm. Ag_2S plays a role in inducing thermal stability in ZnS/Ag_2S bimetallic nanoparticles. The coalescence of nanoparticles for the degradation of pollutants efficiency of more than 70%. Therefore, it is possible to use such a synthetic material as a substitute for a commercial photocatalyst that is sustainable and economical for the degradation of environmental pollutants [40].

The synthesized ZnO-ZnS/Ag_2O-Ag_2S nanocomposites were used for photocatalytic degradation of 2-chlorophenol (2-CP) and complex organic compounds in practical mud. The photocatalytic degradation of 2-CP is related to the pH of the solution, the concentration of the original pollutants and the reaction time. Meanwhile, the adsorption of 2-CP also plays a very important role in its explanation. The maximum interaction and degradation of 2-CP could be found at pH = 5. When the initial concentration increased from 10 to 100 mg/L, the photocatalytic degradation of 2-CP decreased from 96 to 55%, and the best concentration of pollutants was 30 mg/L. High adsorption and photocatalytic degradation of 2-CP were inspected with a maximum removal of 89%. The ZnO-ZnS/Ag_2O-Ag_2S photocatalyst shows a great increase in wastewater solubility, while soluble chemical oxygen demand (COD) increases more than twice its original concentration. It can be found that the synthesized nanomaterials are effective catalysts for many applications, such as the degradation of toxic pollutants, the solubilization of wastewater [41].

ZnO/ZrO_2 nanocomposites with different ZnO: ZrO_2 molar ratios (2:1, 1:1 and 1:2) were prepared by sol gel approach under ultrasonic irradiation. To nanocomposite, the molar ratio of ZnO with different gel incorporated directly into the gel ZrO_2. The reaction mixture was continuously mixed for two days, and then sonicated for 30 min. The filtered composite gel was washed and then calcined in a furnace at 300 °C for 3 h. The photocatalytic activity comparison between nanocomposite and nano-ZnO showed that the photodegradation performance of ZrO_2: ZnO with 1:2 molar ratio was the highest. In addition, the ultrasonic irradiation during the synthesis of nanocomposites results in the enhancement of the photocatalytic activity of the nanocomposites. The pH and the concentration of contaminants in the solution also have some effects on the photocatalytic properties of the nanocomposites [42].

7.3.2 Bismuth-Based Photocatalysts

A kind of Ni-doped $BiVO_4$ photocatalyst, which can be used under visible light irradiation, was synthesized by microwave hydrothermal method. When the doping amount of Ni is 1 wt%, most of the water pollutants such as ibuprofen, Escherichia coli (bacteria) and green tide (phytoplankton) have a good degradation. All percentages of Ni-doped $BiVO_4$ samples have better properties than pure $BiVO_4$. In 90 min, the degradation rate of ibuprofen can reach 80%, the degradation of E. coli can reach 92% in 5 h, and the inactivation of green tide can reach 70% under the irradiation of 60% visible light. The sample shows Ni doping in the vanadium site to have the most stable configuration and also interstitial energy states and oxygen vacancies. These service as electron trapping centers, reducing the migration time of photogenerated carriers and increasing the separation efficiency of electron-hole pairs, which is why this photocatalyst is highly efficient and has antibacterial and anti-algae activity. These properties also show that Ni-doped $BiVO_4$ can be a potential multifunctional material in wastewater treatment [43].

There is a mild and economic method to fabricate Sn-doped BiOCl. This photocatalyst has good photocatalytic activity for degrading dye wastewater under natural sunlight. By controlling the pH of SnO doped BiOCl, the different morphologies of the synthesized catalyst is attained since chemical structures including photocatalytic properties are not the same with different pH. The catalyst showed the best photodegradation performance when mixed dye wastewater (rhodamine B, methyl orange, orange IV and malachite green) was studied as the target pollutant [44].

References

1. X. Duan, H. Sun, Y. Wang, J. Kang, S. Wang, N-doping-induced nonradical reaction on single-walled carbon nanotubes for catalytic phenol oxidation. ACS Catal. **5**, 553–559 (2015)
2. V. Maroga Mboula, V. Héqueta, Y. Andrès, Y. Gru, R. Colin, Assessment of the efficiency of photocatalysis on tetracycline biodegradation. Appl. Catal. B: Environ. **162**, 437–444 (2015)
3. N.G. Moustakas, F.K. Katsaros, A.G. Kontos, G. Em Romanos, D.D. Dionysiou, P. Falaras, Visible light active titanium dioxide photocatalytic filtration membranes with improved permeability and low energy consumption. Catal. Today **224**, 56–69 (2014)
4. T.E. Agustina, H.M. Ang, V.K. Vareek, A review of synergistic effect of photocatalysis and ozonation on wastewater treatment. J. Photochem. Photobiol. C: Photochem. Rev. **6**, 264–273 (2005)
5. U. Černigoj, U.L. Štangar, P. Trebše, Photocatalytic titanium dioxide coatings: Effect of substrate and template. Appl. Catal. B: Environ. **75**, 229–238 (2007)
6. M. Mehrjouei, S. Müller, D. Möller, Design and characterization of a multiphase annular falling-film reactor for water treatment. Chem. Eng. J. **263**, 209–219 (2015)
7. L.S. Li, W.P. Zhu, P.Y. Zhang, Z.Y. Chen, Photocatalytic oxidation and ozonation of catechol over carbon-black-modified nano-titanium dioxide thin films supported on Al sheet, W.Y. Han. Water Res. **37**, 3646–3651 (2003)
8. R.R. Giri, H. Ozaki, T. Ishida, Synergy of ozonation and photocatalysis to mineralize low concentration 2,4-dichlorophenoxyacetic acid in aqueous solution. Chemosphere **66**, 1610–1617 (2007)

References

9. M. Mehrjouei, S. Müller, D. Möller, Degradation of oxalic acid in a photocatalytic ozonation system by means of Pilkington Active™ glass. J. Photochem. Photobiol. A: Chem. **217**, 417–424 (2011)
10. G.Z. Liao, D.Y. Zhu, L.S. Li, B.Y. Lan, Enhanced photocatalytic ozonation of organics by g-C_3N_4 under visible light irradiation. J. Hazard. Mater. **280**, 531–535 (2014)
11. J.D. Xiao, Y.B. Xie, F. Nawaz, Y.X. Wang, P.H. Du, H.B. Cao, Preparation of short, robust and highly ordered titanium dioxide nanotube arrays and their applications as electrode. Appl. Catal. B: Environ. **183**, 417–425 (2016)
12. Y. Ling, G.Z. Liao, Y.H. Xie, J. Yin, J.Y. Huang, W.H. Feng, L.S. Li, Degradation and inactivation of tetracycline by titanium dioxide photocatalysis. J. Photochem. Photobiol. A: Chem. **329**, 280–286 (2016)
13. M.M. Ye, Z.L. Chen, X.W. Liu, Y. Ben, J.M. Shen, Ozone enhanced activity of aqueous titanium dioxide suspensions for photodegradation of 4-chloronitrobenzene. J. Hazard. Mater. **167**, 1021–1027 (2009)
14. Y. Jing, L.S. Li, Q.Y. Zhang, P. Lu, P.H. Liu, X.H. Lü, Photocatalytic ozonation of dimethyl phthalate with titanium dioxide prepared by a hydrothermal method. J. Hazard. Mater. **189**, 40–47 (2011)
15. H. Ghouas, B. Haddou, M. Kameche, Z. Derriche, C. Gourdon, Extraction of humic acid by coacervate: investigation of direct and back processes. J. Hazard. Mater. **205–206**, 171–178 (2012)
16. L.B. Reutergådh, M. Iangphasuk, Photocatalytic decolourization of reactive azo dye: A comparison between titanium dioxide and us photocatalysis. Chemosphere **35**, 585–596 (1997)
17. H. Hao, J. Zhang, The study of Iron (III) and nitrogen co-doped mesoporous titanium dioxide photocatalysts: synthesis, characterization and activity. Microporous Mesoporous Mater. **121**, 52–57 (2009)
18. Y. Xu, Y.P. Mo, J. Tian, P. Wang, H.G. Yu, J.G. Yu, The synergistic effect of graphitic N and pyrrolic N for the enhanced photocatalytic performance of nitrogen-doped graphene/titanium dioxide nanocomposites. Appl. Catal. B: Environ. **181**, 810–817 (2016)
19. P. Wang, J. Wang, X.F. Wang, H.G. Yu, J.G. Yu, M. Lei, Y.G. Wang, One-step synthesis of easy-recycling titanium dioxide-rGO nanocomposite photocatalysts with enhanced photocatalytic activity. Appl. Catal. B: Environ. **132–133**, 452–459 (2013)
20. Y.N. Huo, X.F. Chen, J. Zhang, G.F. Pan, J.P. Jia, H.X. Li, A highly sensitive electrochemical sensor for nitrite detection based on Fe_2O_3 nanoparticles decorated reduced graphene oxide nanosheets. Appl. Catal. B: Environ. **148–149**, 550–556 (2014)
21. G. Marci, V. Augugliaro, M.J. López-Munoz, C. Martin, L. Palmisano, V. Rives, M. Schiavello, R.J.D. Tilley, A.M. Venezia, Synthesis of titanium dioxide via hydrolysis of titanium tetraisopropoxide and its photocatalytic activity on a suspended mixture with activated carbon in the degradation of 2-naphthol. J. Phys. Chem. B **105**, 1033–1040 (2001)
22. J.L. Li, L. Liu, Y. Yu, Y.W. Tang, H.L. Li, F.P. Du, Preparation of highly photocatalytic active nano-size titanium dioxide-Cu_2O particle composites with a novel electrochemical method. Electrochem. Commun. **6**, 940–943 (2004)
23. X.Z. Li, F.B. Li, C.L. Yang, W.K. Ge, Photocatalytic degradation of 2-phenylphenol on titanium dioxide and ZnO in aqueous suspensions. J. Photochem. Photobiol. A: Chem. **141**, 209–217 (2001)
24. Y.R. Do, W. Lee, K. Dwight, A. Wold, MoO_3 in self-organized titanium dioxide nanotubes for enhanced photocatalytic activity. J. Solid State Chem. **108**, 198–201 (1994)
25. X.Z. Fu, L.A. Clark, Q. Yang, Enhanced photocatalytic performance of titania-based, M.A. Anderson. Environ. Sci. Technol. **30**, 647–653 (1996)
26. Y.K. Takahashi, P. Ngaotrakanwiwat, Energy storage titanium dioxide-MoO_3 photocatalysts, T. Tatsuma. Electrochim. Acta **49**, 2025–2029 (2004)
27. J. Papp, S. Soled, K. Dwight, Surface acidity and photocatalytic activity of titanium dioxide, WO_3/titanium dioxide, photocatalysts, A. Wold. Chem. Mater. **6**, 496–500 (1994)
28. H. Zhao, Y.M. Dong, P.P. Jiang, G.L. Wang, J.J. Zhang, C. Zhang, $ZnAl_2O_4$ as a novel high-surface-area ozonation catalyst: one-step green synthesis, catalytic performance and mechanism. Chem. Eng. J. **260**, 623–630 (2015)

29. P. Niu, J. Hao, Efficient degradation of organic dyes by titanium dioxide–silicotungstic acid nanocomposite films: influence of inorganic salts and surfactants. Colloids Surfaces A **443**, 501–507 (2014)
30. M.A. Rauf et al., An overview on the photocatalytic degradation of azo dyes in the presence of TiO_2 doped with selective transition metals. Desalination **276**, 13–27 (2011)
31. M. Malika et al., Evaluation of bimetal doped TiO_2 in dye fragmentation and its comparison to mono-metal doped and bare catalysts. Appl. Surf. Sci. **368**, 316–324 (2016)
32. F. Xu et al., Investigation of titanium dioxide/ tungstic acid -based photocatalyst for human excrement wastewater treatment. Acta Astronaut. **146**, 7–14 (2018)
33. M. Mashkour et al., Catalytic performance of nano-hybrid graphene and titanium dioxide modified cathodes fabricated with facile and green technique in microbial fuel cell. Prog. Nat. Sci. Mat. Int. **27**, 647–651 (2017)
34. G.P. Fotou, S.E. Pratsinis, Photocatalytic destruction of phenol and salicylic acid with aerosol made and commercial titania powders. Chem. Eng. Commun. **151**(1), 251–269 (1996)
35. H. Abdullah et al., Modified TiO_2 photocatalyst for CO_2 photocatalytic reduction: an overview. J. CO2 Utilization **22**, 15–32 (2017)
36. N. Couselo, F.S. García Einschlag, R.J. Candal, M. Jobbágy, Tungsten-doped titanium dioxide vs pure titanium dioxide photocatalysts: effects on photobleaching kinetics and mechanism. J. Phys. Chem. C **112**, 1094–1100 (2008)
37. S. Singh, P.K. Singh, H. Mahalingam, Novel floating Ag+ doped titanium dioxide/polystyrene photocatalysts for the treatment of dye wastewater. Ind. Eng. Chem. Res. **53**, 16332–16340 (2014)
38. T. Ozge, I. Hatice, D. Anatoli, The leaching kinetics and mechanism of potassium from phosphorus-potassium associated ore in hydrochloric acid at low temperature. Sep. Sci. Technol. **52**, 778–786 (2017)
39. H. Bashiri, Cu@SnS/SnO_2 nanoparticles as novel sorbent for dispersive micro solid phase extraction of atorvastatin in human plasma and urine samples by high-performance liquid chromatography with UV detection: Application of central composite design (CCD), M. Rafiee. Ultrason. Sonochem. **36**, 517–526 (2017)
40. M. Abbasi, U. Rafique, G. Murtaza, M.A. Ashraf, Synthesis, characterisation and photocatalytic performance of ZnS coupled Ag_2S Nanoparticles. Arab. J. Chem. (2018)
41. M. Anjum, R. Kumar, M.A. Barakat, Visible light driven photocatalytic degradation of organic pollutants in wastewater and real sludge using ZnO–ZnS/Ag_2O–Ag_2S nanocomposite. J. Taiwan Inst. Chem. Eng. **77**, 227–235 (2017)
42. S. Aghabeygi, M. Khademi-Shamami, ZnO/ZrO_2 nanocomposite: sonosynthesis, characterization and its application for wastewater treatment. Ultrason. Sonochem. **41**, 458–465 (2018)
43. C. Regmi, Y.K. Kshetri, T.H. Kim, R.P. Pandey, S.K. Ray, S.W. Lee, Fabrication of Ni-doped $BiVO_4$ semiconductors with enhanced visible-light photocatalytic performances for wastewater treatment. Appl. Surf. Sci. **413**, 253–265 (2017)
44. X. Han, S. Dong, C. Yu, Y. Wang, K. Yang, J. Sun, Controllable synthesis of Sn-doped BiOCl for efficient photocatalytic degradation of mixed-dye wastewater under natural sunlight irradiation. J. Alloy. Compd. **685**, 997–1007 (2016)

Index

A

Ammonium metavanadate, 14
Anatase titanium dioxide, 15, 18, 21, 38, 48, 65, 66, 75, 96, 103
Artificial photosynthesis, 2
Autoclave, 14, 56

B

Bandgap, 21, 41, 46, 130
Beer-Lambert's law, 6
Binding energies, 24, 32, 59
BiOX, 47, 48
Bismuth-based photocatalyst, 47, 65, 148
$BiVO_4$, 49, 75, 148
Brunauer-Emmett-Teller (BET), 13, 45
Bulk-charge separation, 9

C

Calcination, 14–16, 26, 28, 29, 48, 105, 106
Carbon decorated, 15, 34, 65, 74, 103, 116
Carbon spheres, 14, 26, 30, 32, 58, 65, 73, 103, 114
C-doped titanium dioxide, 73
Charge carrier transfer, 69, 72
Chemical vapor deposition, 45, 46, 88
Chromophore, 7
Coal-fired flue gas, 66, 103, 134
Conduction band, 4, 9, 20, 70, 74, 86, 95, 107, 110, 111, 115, 120, 127, 133, 137
Convex mechanism, 112
Core-shell nanocomposites, 30
Crystalline size, 29
$CSs-BiOI/BiOIO_3$, 45, 56, 57, 60, 129, 131, 132
$Cu(OH)_2$, 14, 26
$Cu(OH)_2$/titanium dioxide@C, 14, 26
CuO/titanium dioxide, 14, 26–31, 34, 35, 65, 70, 73, 103, 114

D

Decompose, 2, 4, 87, 136
Doping, 14, 16, 19, 26, 29–33, 36, 40, 46, 48, 61, 62, 65–70, 73–75, 106, 107, 113–120, 129, 130, 133, 134, 143, 145, 148

E

EDAX, 26, 27
EDX, 46
Electron and hole recombination, 33
Electron-hole pairs, 23, 47, 66, 70, 72, 74, 75, 85, 113, 114, 122, 129, 148
Energy-dispersive x-ray spectroscopy, 18
Environmental protection, 2, 84, 85, 88

F

Facets, 48, 49, 53, 66, 75, 103, 127
Fe_3O_4-titanium dioxide, 83, 89–91, 93, 94
FESEM, 26, 27, 46
Flower-like, 45–47, 89
Flue gas, 66, 83–85, 88, 89, 95, 97, 98, 103, 108, 132, 141
FTIR, 13, 23, 24, 41, 60, 94, 95
Functional groups, 23, 31

G

Grotthuss-Draper law, 6

H

Heterogeneous photocatalysis, 4

Homojunction, 21, 107
HRTEM, 13, 18, 19, 38, 54, 55, 59, 89
Hydrothermal, 3, 14, 26, 45–49, 57, 58, 66, 83, 89, 94, 148
Hydroxyl groups, 31, 36, 48, 61, 123, 142
Hydroxyl radicals, 4, 31, 134

I
In_2O_3/titanium dioxide, 15, 36, 38, 41, 65, 74, 103, 116, 119, 121, 124
Interplanar spacing, 18
Intrinsic oxygen vacancy, 48

L
Lattice oxygen, 24, 36, 69, 123
Lattice spacing, 18

M
Mercury, 66, 87, 103, 104, 113–115, 123, 125, 129, 131, 134
Metal modified titanium dioxide, 65
Mineralization, 47, 141, 142
Modified photocatalyst, 15, 65, 143
Morphology, 13, 16, 18, 26, 36, 45, 47, 48, 58, 59, 65, 103
Morse potential energy curve, 7

N
N_2 adsorption-desorption isotherm, 18, 21, 40, 136
Nanocube, 36
Nanoparticle, 3, 4, 16, 18, 31, 36, 38, 57, 58, 61, 66, 73, 76, 88, 118, 147
Nanostructure, 46, 72
NaOH, 14, 46
NHE, 74
Nonmetal modified titanium dioxide, 65
NO_x, 83–86, 92, 95, 96, 99

O
One-step hydrothermal method, 67
Optical absorption threshold, 4
Oxidation capacity, 115, 130
Oxidation reaction, 24, 49, 75, 128

P
P_{25}, 14, 16, 21, 23, 26, 29–33, 36, 40, 41, 106, 107, 115–117, 145, 146
Photocatalytic activities, 47, 48, 66, 69, 125
Photocatalytic performance, 33, 45, 47, 48, 65–67, 69, 73, 75, 76, 107, 112, 117, 118, 134, 135
Photocatalytic reaction, 5, 8, 19, 24, 39, 40, 52, 87, 113, 120, 127, 133, 136

Photocatalytic reaction mechanism, 5, 123, 127, 133
Photocatalytic technology, 2, 85, 87, 141
Photochemistry, 6
Photodegradation, 47, 69, 73, 143, 147, 148
Photoexcited electrons and holes, 23, 73, 74, 116, 118
Photoluminescence, 13, 22, 23, 33, 45, 52, 116, 117
Photon(s), 4, 5, 7, 50
Pollutants, 2–4, 66, 76, 83–85, 88, 141, 147
Pore diameter, 31, 130
Pore volume, 31, 130
Power plant, 84, 97, 99, 105, 108

Q
Quantum, 5, 67, 70, 85–88, 136, 143
Quantum efficiency, 143

R
Red shift, 61, 62, 67, 134
Removal efficiency, 31, 53, 91, 93, 95, 97, 114, 117, 119, 120, 126, 129, 131, 136, 142

S
Scanning electron microscope, 13, 36, 45
Semiconductor, 1–4, 8, 9, 33, 45–48, 50, 51, 66, 70, 72–75, 85–87, 110, 127, 136, 142, 143
SO_2, 83, 84, 90, 91, 93–96, 108, 110
Solar light, 47
Solvothermal, 45, 47
Specific surface area, 13, 19, 39, 47, 52, 65, 76, 87, 88, 98, 107, 133, 136, 143
Stark-Einstein law, 6
Stretching vibration, 22, 23, 31, 41, 94
Superoxide anion, 4

T
Tetrabutyl titanate, 89
Titanium dioxide, 1–4, 13–16, 18, 19, 24, 26–33, 35, 36, 38, 40, 41, 48, 65, 66, 70–76, 83, 85, 87–89, 91–96, 98–100, 103, 105, 107, 114–123, 125, 142–146
Transition metal, 66, 67
Transmission electron microscopy, 13, 18, 37, 38, 45

U
Ultraviolet, 4, 5, 19, 70, 74, 75, 85, 87, 114
UV radiation, 5
UV-vis absorbance spectra, 22, 30, 35
UV-Vis DRS, 13, 19, 30, 51, 118, 134

V

V_2O_5/titanium dioxide, 14, 15, 65, 72, 99, 103, 105

Valence band, 4, 9, 48–50, 59, 70, 75, 86, 95, 107, 110, 111, 115, 120, 121, 129, 130, 133, 134

Vibronic transition of electron, 7

Visible light, 3, 5, 7, 19, 20, 22, 31, 33, 47, 48, 50, 60, 63, 66–68, 70–74, 86, 87, 105–108, 110, 111, 116–121, 123, 125, 127, 134, 143, 148

W

Waterfall mechanism, 112

Wavelength, 4–6, 29, 46, 50, 52, 62, 87, 105, 125, 129, 130

X

X ray Photoelectron Spectroscopy (XPS), 13, 24, 25, 31, 34, 35, 37, 46, 54, 56, 59, 61, 65, 94–96, 108, 109, 111, 113, 120, 121, 123, 124

X-ray diffraction, 13, 15, 45, 134, 135

Z

Zinc-based photocatalyst, 45, 147

$Zn(NO_3)_2 \cdot 6H_2O$, 46

ZnO, 45, 46, 66–72, 86, 103, 134, 136, 137, 142, 147

Printed by Printforce, the Netherlands